RABBITS

RABBITS
A GUIDE TO MANAGEMENT

JOHN SANDFORD

The Crowood Press

First published in 1988 by
The Crowood Press
Ramsbury, Marlborough,
Wiltshire SN8 2HE

© John Sandford 1988

All rights reserved. No part of this publication may be reproduced or transmitted in any form or by any means, electronic or mechanical, including photocopy, recording, or any information storage and retrieval system without permission in writing from the publishers.

British Library Cataloguing in Publication Data

Sandford, John
   Rabbits.
   1. Livestock. Rabbits. Care and
   management – Manuals
   I. Title
   636'.9322

ISBN 1–85223–072–X

Acknowledgements

The author is most grateful for the help of Roger Parkin in reading the manuscript, to all those who have helped in many other ways and to Bill Thomas for the use of photographs on pages 12, 13 and 39.

Line illustrations by Claire Upsdale-Jones

Typeset by Inforum Ltd, Portsmouth
Printed in Great Britain by Butler & Tanner Ltd, Frome and London

Contents

	Introduction	7
1	Housing and Equipment	10
2	Breeds and Breeding	39
3	Nutrition and Feeding	49
4	Management and Stockmanship	59
5	Other Aspects of the Rabbit	84
6	Disease	89
7	Marketing and the Economics of Rabbit Farming	103
8	Records and Recording	108
	Further Reading	121
	Useful addresses	123
	Index	125

Introduction

Rabbits have been maintained in captivity, under one system or another, for many hundreds of years. The animal itself almost certainly originated in what is now Northern Africa and the Iberian Peninsula – the Romans took it back from there to Italy, where they kept it in enclosures, called *leporaria*. Its flesh was considered by them to be a great delicacy, and the medievel monks some centuries later evidently had the same tastes.

Throughout the Middle Ages the 'right of warren', or the right to keep rabbits in a more or less wild state but nevertheless confined, belonged solely to the king. On occasions he would grant it to others as a favour.

During the same period, there was a considerable increase in the keeping of rabbits for food by monasteries – the monks had adopted the rather novel interpretation that unborn and new-born rabbits were not meat, and could therefore be eaten on Fridays!

Warren rabbit keeping survives to the present day, although not in any sense in the same way as it did in the past. True farming, with closely-confined rabbits bred on what amounted to an intensive scale, was established as long ago as the beginning of the nineteenth century. There are even older records of the tiering of cages for rabbit keeping, but it was the early eighteen hundreds that saw the setting-up of large farms with rabbits in extremely densely-stocked houses.

Throughout Europe, there has always been a pattern of small-scale rabbit keeping for meat. It was unusual for more than ten or fifteen breeding does to be kept and these would probably not have produced more than some twenty or twenty-five young per year. All the animals were fed with freely and cheaply available roughages and hay, and feeding was thus a very low cost activity.

There have been a number of attempts at different times in the past hundred years to establish not only meat but also fur and wool farming. None of these was permanently successful. After the 1914–18 war, it seemed that an interesting fur farming industry, based on the domestic rabbit, might be set up, and some quite large farms did achieve an amount of success. However, enormous imports of cheap fur in the early 1930s ruined the prospects of this business. Similarly, the Angora rabbit wool industry looked promising but, again, excessively cheap

Introduction

imports (ironically, from rabbits sent out from the UK in the first place) destroyed the market.

In 1953, myxomatosis was introduced to the UK and to the whole of Europe. The annual crop of some 50 million or more wild rabbits was lost. These animals had supplied meat, and fur for the manufacture of hats (incidentally, a good bowler hat could only be made from rabbit fur) and in the next few years the possible replacement of this large market by meat and fur from domestic rabbits was widely considered. It is surprising to think nowadays, when meat is really the only saleable commodity, that this thinking centred mainly on finding a substitute for the fur of the wild rabbit. As myxomatosis was present on the Continent too, the enormous supply of domestically-grown skins, particularly from France, had diminished at the same time. On the other side of the world, the Australians had at last achieved their dream of introducing myxomatosis and the wild rabbit population had been decimated – Australia was, therefore, in the same situation as the European countries.

So, by the late 1950s a domestic rabbit industry of some kind was being established and it was quite different from anything which had been done before. This new industry was based almost entirely on the original American idea of single tiered metal cages (which, it was suggested, were totally self-cleaning), coupled with the feeding of a complete pelleted ration and the production of large litters of rabbits with fast growth, for slaughter at ten weeks of age or so. Such a system necessitated both a high fixed asset basis and a sufficiently large working capital. The main aim was to keep labour costs to a minimum.

The same situation began to emerge in Europe. Previously there had been quite a large industry based on very small units with solid floor cages and usually farm or collected foods, but within ten or twelve years of the development of the new system, production had soared. In France alone it reached 300,000 tons (305,000 tonnes). From 1970 to 1980 the number of producers dropped from one million to 650,000, but the average size of a unit with more than twenty does almost doubled in the same period.

France, Germany, Italy and (more recently) Spain have all travelled the same path. Unfortunately, this has not been the case in the UK, where the early promise of the industry at the beginning of the 1960s has not been achieved. Any number of reasons are put forward, but it is a fact that in many cases the wrong type of person entered the industry, usually with totally insufficient capital and with no knowledge or experience of rabbit husbandry, or, indeed, of the raising of any form of livestock. Also there were (and unfortunately still are) the unscrupulous people who saw that they might make a quick profit by misleading

Introduction

inexperienced or gullible folk and selling them the poorest of stock, which, even with the best possible equipment and skill, could never make a profit. Needless to say, after a short time the newcomers were forced to retire, certainly poorer, and often not realising why their enterprise had failed.

The chaotic state of the marketing of the produce added to these problems. The success of any business depends to a large extent on the success of its marketing. In the case of the rabbit industry, whilst there is a core of good, reliable processors to buy the meat animals, there are many more who come and go, often causing financial problems to the rabbit farmer. Happily, there are some signs that changes may take place in the future, and this should help to alleviate the problem.

The formation of the Commercial Rabbit Association (CRA), in 1960, was carried out in an effort to bring together all those with an interest in the rabbit industry. At the time, it was hoped that such an organisation would prevent the problems outlined, or at least try to help the industry to steer clear of them. The CRA has had a chequered career but now seems to be on a more stable course. It is an organisation which should be approached by anyone thinking of entering the field, and this should be done before the final planning stage and certainly before any money is invested in an enterprise. It is also an organisation which should be supported by everyone who has an interest in commercial rabbit farming.

The commercial rabbit industry, then, is based on the keeping of highly-productive does in metal cages (almost invariably single tier), for the production of young meat rabbits to be sold at around ten weeks of age – these will weigh between four and five pounds and are almost invariably sold live to a processor. Both the feeding (in pellet form) and the management are highly intensive.

It cannot be denied that there are some rabbit farmers who enjoy a satisfactory return for their investment and labour. It is, however, equally true that there are many more who, at best, break even and, at worst, make a serious loss. The main problem lies in a lack of knowledge. Rabbit farming is certainly no easier than any other form of livestock husbandry and, indeed, in many ways it is more difficult. This fact is not generally recognised. Large numbers of people who would never dream of keeping any other livestock persuade themselves that rabbits are easy to look after. They are encouraged by the thought that many children are able successfully to keep rabbits as pets, but they forget that farming commercial rabbits, and making a profit from them, is an entirely different and more complicated proposition. This book is an attempt to explain the basic facts that any aspiring rabbit farmer will need to know.

1 Housing and Equipment

There are seven separate aspects to the housing of rabbits in the broadest sense that need to be considered.

1. The site.
2. The building (if any) or the protection of cages.
3. The control of the environment.
4. Caging.
5. Watering systems.
6. Other equipment.
7. The cleaning and maintenance of housing and equipment.

A range of rabbit houses.

Housing and Equipment

THE SITE

It is fairly rare in the UK (although much more common on the Continent) that a properly designed, purpose-built house is available for the rabbit enterprise. Most rabbits are housed in buildings which were previously used for other purposes. These range from old poultry houses, barns, stables and even old pigsties. In the same way, the location and site are almost invariably fixed.

The important points which need to be looked at in relation to the site, however, include the following:

1. Protection from high winds and very inclement weather.
2. Reliable supplies of both water and electricity.
3. Good access for heavy lorries.
4. An area away from rabbit houses for the disposal of manure.
5. Fencing to keep out intruders, both predators and human.
6. Good drainage to prevent waterlogging of the site.
7. Sufficient room for future expansion.

Although modern commercial rabbit farming is highly intensive, the area requirements are usually underestimated. A single doe unit on the flat deck system must include some 8sq ft (0.75sq m) for the doe and young to five weeks of age (plus 2.2sq ft (0.2sq m) if the young are not weaned by five weeks), a proportion for bucks and replacement animals, passage ways, room for storage and so on, and lastly cage space for fattening young. In all, an area approaching 21sq ft (2sq m) is necessary for each breeding doe and her followers.

The site area is also usually underestimated. It is safest to allow at least twice the final building area for the minimum size plus land for future expansion. It is a serious error to establish a rabbit unit on a site which allows no room for future expansion.

RABBIT HOUSES AND SHELTERS

Apart from a few special cases where brick houses were built for the rabbit cages, until quite recently it was unusual to give cages more than an extended roof as protection. This applied mostly to three tier cages built in stacks. Sometimes, with cages facing each other, the roof of each stack was extended slightly more and thus overhead protection was obtained. This system has been modified to a simple shelter in which wire cages can be suspended, but this is only possible where temperatures do not fall much, if at all, below freezing point, and then only

Housing and Equipment

occasionally. Examples can often be found in Spain and in parts of America.

A further development in recent years is the tunnel house which was developed in an attempt to reduce the high cost of houses for rabbit units. It essentially consists of a metal framework over which plastic film and, in some cases, netting, is fixed. Various types of insulation can be built in and some openings are usually left along the lower side walls for ventilation.

The covering of the tunnel house is often a sandwich construction of two layers of plastic with a heat insulating layer between. A building 12ft (3.6m) in width with 8ft (2.5m) to the ridge is very suitable and the building is usually built up in sections of 8ft.

The advantages of the tunnel building are:

- It is relatively inexpensive.
- It usually has excellent natural ventilation.
- It can be moved with relative ease.
- Environment control costs are eliminated.
- Size can be increased at any time with ease.

Tunnel house, exterior view.

Housing and Equipment

Tunnel house, interior view.

Disadvantages are:

- It is more difficult to keep out vermin.
- It is not so permanent.
- It probably requires a higher level of management.
- Food costs in winter are increased slightly.
- Problems arise with freezing of the water system in bad weather.
- It has been found impossible to achieve the very high production obtained in some fully environmentally-controlled houses.

The use of shelters is not really possible when intensive husbandry is contemplated. The high cost of building new housing has meant that in very many cases old buildings have to be converted. It must be realised that converted buildings can never be as satisfactory as buildings designed for the purpose, and if it is to be done properly conversion is, in most cases, quite expensive. A further problem which is likely to arise with converted buildings is that there is usually no room for expansion if the enterprise proves a success.

The requirements of the rabbit house are that it should be of a suitable length, width and height to provide sufficient space for the number of

Housing and Equipment

A small purpose-built rabbit house with single tier staging mounted on concrete blocks – not the most satisfactory arrangement.

breeding does which it is to accommodate, and that it should be built of materials which compromise well on cost and length of life, insulation characteristics and serviceability (capacity to be kept clean and free of damage and vermin and so on).

There is such a wide range of possibilities in building rabbit houses that it is impossible to point out more than a few important planning criteria.

1. The width should allow the correct multiple of cages with adequate passage ways in between, with no cages nearer than 8in (20cm) to the wall, and the passage ways 3ft (1m) or more of clear space.
2. The total length must be a multiple of the length or depth of the cages as arranged.
3. The height to the eaves should give an ideal total density of liveweight of between 3 and 5kg per cubic metre, always providing the ventilation is satisfactory.

Housing and Equipment

4. The walls should be as smooth as possible and the number of ledges kept to a minimum. Light and ventilation can be provided as well without windows as with, in some cases more easily. The only point is that windows should be fine-netted to prevent the entry of flies and other vermin.
5. Doors should be well fitting and doorways sufficiently large to take the widest trailer that is likely to be used.
6. The floor is very important. Undoubtedly the best type is a well-made and finished concrete floor with suitable guttering to allow for drainage. All other types of floor are far less satisfactory. Should simple earth floors be used (possibly in extended shelters or tunnel houses) then the very least requirement is a good concrete passage way between cages.

CONTROL OF THE ENVIRONMENT

The reasoning that lies behind the environmentally-controlled house is that to achieve its fullest production potential the rabbit requires ideal surroundings as well as ideal food and management. Although there are some breeders who get extremely good production figures without the use of such buildings, there are none which get results similar to those of the best ten per cent or so of farmers on the Continent, who all rely on full environmental control. They achieve rates of production of over seventy young per breeding cage per year, weighing over 260lb (120kg) with a food conversion rate of under four. These figures are quite exceptional compared to those obtained under other circumstances which may average less than half the number of young, weighing half as much and with a worse food conversion rate.

Ventilation in the rabbit house is of prime importance, not only in fully-controlled houses, but also in any building in which rabbits are kept. There is no doubt at all that there are a number of diseases which are started or aggravated by poor ventilation. Ventilation in the more heavily-stocked environmentally-controlled houses can be a complicated matter – not only does there have to be a suitable rate of air change to eliminate the excess moisture, dust and gasses in the air, but this air change must assist or match the temperature control and must be of such speed and direction that it does not in itself cause harm to the animals.

In the cases where natural ventilation is to be used, it is important that there should be some means of adjusting the size of the inlets and outlets. The total inlet area should, in mid opening, be about 12in^2 (75cm^2) for every 2lb (1kg) of liveweight of rabbits that the house is

Housing and Equipment

likely to contain and the total outlet area should be greater than the inlet area. Whether it is better to have the outlets high and the inlets low, or vice versa, is the subject of much argument, but what is certain is that the direction of air movement should be checked to ensure that draughts are not directed on to the animals.

In environmentally-controlled houses, the aim is to regulate the temperature to 60–68°F (15–20°C). The optimum temperature is probably 60°F (16°C). Excessive temperature can be reduced by sucking air through a wet mat, but this increases humidity. Excess temperature is as bad as, if not worse than, cold.

It is often overlooked that the frequency of manure removal has an effect on ventilation requirements. The greater the bulk of manure that is allowed in the house, the greater is the requirement for air changes.

The calculations required for a completely satisfactory environment control system are complicated and the means of carrying out the system are varied and require careful and accurate installation. As the cost is relatively high, it is desirable to engage an expert in the subject. Some assistance in the ventilation of other houses is probably also desirable.

RABBIT CAGES AND PENS

There is a great variety of cages and hutches in which rabbits are kept. The traditional wooden three-tiered cage for use outside a building (possibly with a shelter for further protection) has been used for well over 200 years. Whilst such cages have obviously proved successful in many ways, the development of a commercial industry in which the labour costs assume an ever greater significance, and the increasing awareness of the causes and prevention of disease, has led to the development of other types of caging. Nowadays all the fully intensive farms which are successful use metal mesh cages. Other types are useful for far less intensive purposes.

There are four basic arrangements of cages with floors that are either solid or of mesh. Until some twenty-five years ago, the only floors used were solid, most frequently with sawdust underneath straw as litter. The advantages of this system were that outdoor housing was possible, the system was totally suitable for all different breeds (including very large ones and those with poor furring on hocks) and in winter the litter provided protection and some warmth. These factors were however, outweighed by the disadvantages, which included a greatly increased risk of some diseases (although the outside environment helped in the prevention of others), an increased operating time and considerably

Housing and Equipment

increased cleaning time, the problem of the availability of straw and its disposal after being used as litter.

These disadvantages eventually meant that, except under unusual circumstances, the use of solid floors was discontinued. This in itself led to the virtual stopping of outside rabbit farming.

Types of Caging

Cages within a rabbit house can be one, two or three tier (and in some cases even four), the cages in multiple-tier systems being vertically above each other with the manure disposal by automatic means or by sloping solid floors under wire. There is also the 'Californian system', in which one cage is mounted above, and to the rear of, the first cage, while sliding Californian systems exist in which the rear cage can be pulled forward for servicing.

Modified Californian system with top tier brought closer to the passage way – note the suspension on a free-standing frame.

Housing and Equipment

Flat deck system with external nest boxes mounted on a steel frame. Note the rigid water pipe and breaker tanks and the gully on the floor for drainage.

The most common system, however, is the flat deck system. In this set-up all the cages are on the same level and the advantages of this are:

- The cages are all at an ideal height for servicing.
- Manure disposal is simple in that the faeces will fall straight to the floor, or into the pit or trough, and can be easily removed.
- Installation is simple.

The disadvantage of the flat deck system is that the number of cage units in any house is less than with other systems and hence the capital housing cost per cage is higher. The housing density that can be achieved with three tier caging is considerably greater and hence the capital housing costs tend to be lower. The individual cage costs, however, tend to be higher.

There is no question of there being an enhanced labour demand with

Housing and Equipment

Three tier cages with galvanised trough and moving belt discharging through wall – note the watering system and the hoppers.

tiered systems, and it must also be remembered that with these systems the ventilation requirements become more complex.

Requirements of Cages

The requirements of good caging are that it should:

1. Be structurally sound and well made.
2. Be of the correct dimensions and size for the welfare of the animals and ease of service.
3. Be sanitary and capable of being easily cleaned.
4. Be convenient for the handling of the animals and servicing their requirements.
5. Be escape-proof and so constructed that no animal can hurt itself or others.
6. Provide for good visibility of the animals at all times.

Housing and Equipment

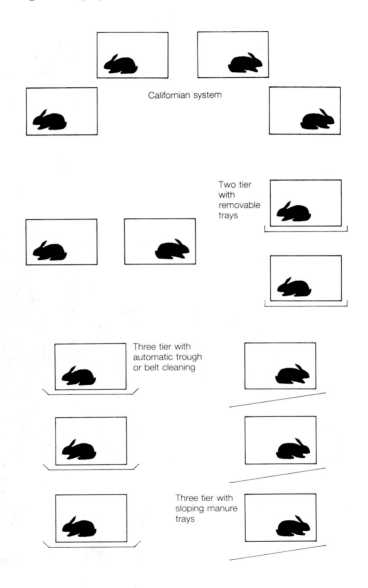

The four typical arrangements of metal cages.

Housing and Equipment

Cage Numbers and Space

The basic unit of caging is the breeding cage. On the Continent, with fairly uniform cage heights of 12in (30cm) and a uniform depth of 20in (50cm), the standard breeding cage (with an external nest box) is 24in (60cm) wide. This is increased to 28–32in (70–80cm) if an internal nest box is used. Future breeding stock is housed in cages of half this width, and bucks have a width of only 16in (40cm). Thus, a continental breeding doe cage (almost invariably with weaning at four weeks) is 3ft^2

Individual flat deck cages mounted on steel framework with manure pit below – note the rigid white water pipe and the very smooth walls.

Housing and Equipment

(0.3m²). In the UK, almost 6ft² (0.56m²) would be the usual recommendation and a standard depth would be 24in (60cm), a standard height would be 16in (40cm) and the width of a breeding cage some 30in (75cm).

With the Ministry of Agriculture and the CRA recently issuing codes of welfare, it is likely that the standard sizes will become larger. The measurements recommended by the Ministry are a minimum height of 18in (45cm), a depth of 24in (60cm) (as before), and a width of about 36in (90cm). The floor space requirements for rabbits are as follows:

- Doe with litter, up to five weeks   6 sq ft (0.56 sq m)
- Doe with litter, up to eight weeks   8 sq ft (0.74 sq m)
- Weaners up to 12 weeks, each rabbit   0.75 sq ft (0.07 sq m)
- Non-breeding rabbits over 12 weeks, per rabbit 2 sq ft (0.19 sq m)
- Adult breeding stock   6 sq ft (0.56 sq m)

The Manufacture of Cages

The material from which cages are made is almost invariably welded wire mesh. Two different types are generally used, one for floors and one for the sides and the roof – the material used for floors can be used all round if necessary. Several types of floor aperture are used with success. For floors, the most useful is probably 0.75in (19mm), but a rectangular mesh up to 3in x 0.5in (75mm x 12.5mm) can also be used. Mesh with apertures larger than 1in x 2in (25mm x 50mm) is unsatisfactory for use at the sides and on the top.

There are three important points to remember with regard to cage construction material. Firstly, the wire of the mesh is frequently too thin for its purpose. Wire is measured by gauge, and for the floors of the cages twelve gauge wire 0.1in (2.5mm) in diameter is desirable. It is possible to use slightly lighter mesh, but it is never as satisfactory. The minimum thickness of the wire in all other applications is fourteen gauge, or 0.08in (2mm) in diameter.

Secondly, the wire mesh should be heavily and well galvanised. As a result of constant use and cleaning, and attack by urine chemicals, galvanising becomes damaged. Some highly successful establishments have their cages regalvanised but this is only worthwhile when the cages are of the best possible quality in the first place.

Thirdly, care must be taken to ensure that the wire mesh itself is free from sharp spikes and tiny metal hooks, caused by solidified molten zinc when the galvanising is unsatisfactorily performed. A rub with the hand will quickly (and painfully) demonstrate when the wire mesh is not up to standard.

Housing and Equipment

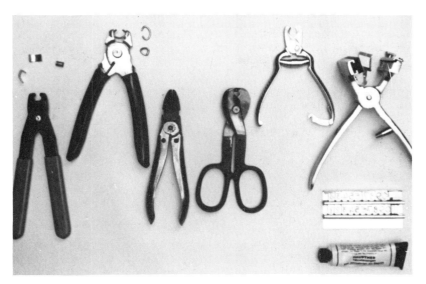

From left to right: Ferrule pliers (with open and closed ferrule above). Clip pliers (with open and closed clips above). Side cutting toggle wire cutters. Ferrule opening (removing) pliers. Claw clippers. Tattooing outfit with tattoo pliers, two sets of numbers and roll-on and tube ink.

Cages can be purchased ready-made or manufactured by the rabbit farmer himself. They can be made from roll wire or from bought panels. In both cases, clip pliers or ferrule pliers must be used to fasten the mesh or the panel together – a common mistake is to put too few ferrules or clips in place. A useful tool is also a type of plier with sharpened points which is used to open and remove the rings or clips and ferrules.

The top of the cage must be so fixed that it will operate as a door through which animals can be removed. Various openings may also have to be cut in the mesh to allow feed hoppers and external nest boxes to be fitted. Whenever openings are cut into the mesh, care must be taken to ensure that sharp edges are removed or, better still, covered.

Installation of Flat Deck Cages

Cages are either suspended from roof members (and it is important to ensure that the roof is strong enough to support the weight which may

Housing and Equipment

A small purpose-built rabbit house with flat deck cages suspended from rafters.

be suspended from it, and to consider the normal safety factors for snow and wind and so on), or supported by walls or other structures such as angle frameworks. The wire suspension is undoubtedly the best, for it avoids any build-up of manure and does not form any interruption to ventilation, both problems that occur with the other systems of support.

There is an alternative system of caging now being introduced, particularly on the Continent. This is the system with very easily-removed cages, which are, of necessity, a standard size and are mounted independently. Any that are damaged or excessively soiled can be removed immediately from the rabbit house for repairs and a thorough cleaning, tasks which can be done with greater ease in a special place, rather than in the rabbit house itself. It is true that the cost of this caging is rather more than that of the standard cages, but in the really professional unit the difference is soon recouped.

Housing and Equipment

Colony Pens

From weaning until two or three weeks before first breeding in the case of does, and from weaning to about twelve weeks in the case of bucks, young rabbits can be kept together in colonies. In all cases, colonies should be strongly made and should not be over-large. Indeed, the practice of using colonies is really suitable only for a large-scale operation where the young are to be kept quite apart from the breeding herd.

A maximum of about forty young rabbits should be placed in each separate colony pen. It is necessary to reduce the number to this level for the greatest success and to prevent the possible disaster of panicking. When rabbits are kept in quantities in large pens, then the space requirements are not as large as when they are more confined. Whilst the space requirements for rabbits up to twelve weeks of age is normally quoted as about fourteen animals to every 10sq ft (1sq m) it is possible to have eighteen or so per square metre in pens which are two square metres or more.

A metal colony pen – note the use of wire hay racks as partitions.

Housing and Equipment

The important point to note in the construction of colony pens is that they must be sufficiently strongly built, with particular attention paid to the support of the floor.

An excellent method of feeding is by the use of large circular hoppers suspended just above the floor.

EQUIPMENT IN THE RABBIT HOUSE

Apart from the cages and the tools for their manufacture, there are other items of equipment which are essential in the rabbit house. Lighting and watering systems, feed hoppers, and nest boxes are the most important items – such things as cleaning equipment, marking equipment, trailers, food scoops and weighing machines will also be needed.

Before discussing these in detail, it should be said that cheap, poorly designed and manufactured equipment is, in the long run, enormously expensive. To the original low cost must be added the cost of labour for maintenance (which will undoubtedly be required), the expense of a loss of production which can often occur, and finally, in the worst cases, the loss of the animals themselves.

Lighting

Little has been scientifically established on the subject of lighting for different groups of rabbits – there is always a great deal of opinion on the subject, much of it contradictory and quite a lot of it wrong. The basic facts that have been established are as follows:

1. Young rabbits do not need much light – a few hours a day are more than sufficient, and that can be at a low intensity. High levels and long periods do not, however, appear to be harmful.
2. Bucks need half the amount of light needed by breeding does. More than eight hours of light reduces slightly the reproductive capacity of bucks. Less than fifteen reduces the performance of does.
3. The intensity of light is more important for the does than for any other group – does need a higher intensity.
4. It is the beginning and ending of light intervals that are important. Intermediate dark periods do not appear to affect the animals.
5. It is probable that the pattern of light rather than the actual periods and intensity may be important.

It follows from this that until more evidence is established, the best arrangement is probably to attempt to shade the bucks slightly and to

Housing and Equipment

give the entire house (if it is mixed) a standard daylight period of fifteen or sixteen hours, controlled by a time switch.

Watering Equipment

Rabbits require a constant source of good, clean water. No rabbit farm will ever have any success unless this exists. A wonderful variety of pots and jars and bottles and troughs is to be found, but no one who is interested in rearing rabbits commercially should ever contemplate anything other than an automatic drinking system. With this, water is piped to each cage and delivered, usually to nipple-type drinkers, although occasionally a bowl, with a valve to regulate the level, is used to hold the drinking water.

The nipple system is without question the best, provided of course, it is sufficiently well made. Cheap nipple drinkers can cause untold trouble, and the amount of water that can be lost from faulty, poorly designed and dirty nipples is remarkably high. This excess water can itself cause problems in the rabbit house. Blocked nipples, with the consequent lack of water to the animals, can cause a considerable loss in growth, loss of production in does, poor health and so on.

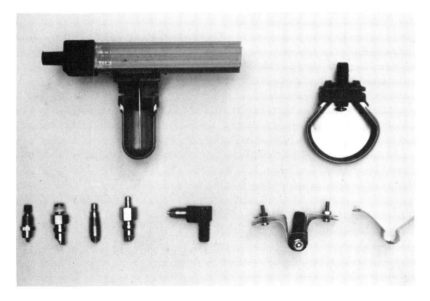

Watering systems components.

Housing and Equipment

Nipple waterers consist of a body which contains a protruding pin with a valve seating (where the nipple connects with the body) on the inside. The body may or may not contain O rings or a filter, and it is flanged or screwed to attach to the piping. Almost invariably the bodies are made of brass and the operating pin of either brass or stainless steel. In some cases the brass pin can be completely ruined by the animals so a stainless steel pin is undoubtedly the best.

The nipples are attached to the water supply in different ways. Sometimes a flexible pipe is used and sometimes a rigid pipe is run the length of the cages. Whilst the rigid pipe is generally more expensive, it is also probably better. The internal diameter of the pipe is usually 0.3in (8mm). This is absolutely the minimum internal diameter that should be used, and in many cases a larger diameter is desirable, particularly with the longer run of pipes. A larger diameter allows a greater flow of water and assists in the prevention of freezing.

The mains water supply is connected (sometimes via a storage tank) to a small low-pressure delivery tank, through a ball valve. The purpose of the large storage tank is to keep a reserve and to allow for the mains water to warm up to room temperature. It is not, however, essential. The height above the line of the nipples at which the delivery tank must be situated depends upon the ideal water pressure of 1–4lb (0.5–1.8kg) – a height of 2.3–9ft (0.7–2.8m) is required, but a level of about 3ft (1m) above the nipples is usually satisfactory. When installing a system, it is certainly best not to fix the tank in its final position until it has been tested. If there is an insufficient flow then the tank must be raised. In some cases, if dripping is being caused by excess pressure, the tank may have to be lowered.

Water coming into the tank should be filtered and a secure cover placed over the tank. In addition it is a good idea to make certain that there is a large plug in the bottom of the tank to allow it to be cleaned out thoroughly from time to time.

When installing a watering system care should be taken to ensure that no dirt gets into it. It is surprising how easily this can happen and it can be a constant source of trouble. Covered vent pipes should be incorporated in long runs of pipe, although it is often better to have separate breaker tanks for shorter runs. Some easy way of shutting off the supply to the tank or tanks should be included, as should some easy way of opening the line at the far end, to allow it to be cleaned and washed through. A totally opaque tubing (black is best) must be used to help prevent any growth of algae in the pipe, which will cause trouble if not discouraged.

Some nipples are manufactured to be operated vertically, some at an angle of 45°, and some work most effectively when placed horizontally. It

Housing and Equipment

A rabbit drinking from a water nipple.

is essential to install each nipple in the correct position and at the correct height. The best height for adult animals is 8–10in (20–25cm) from the cage floor, whilst for very young animals a height of 5–6in (12–15cm) is more appropriate. As a matter of fact, the young will often be seen licking at the wire when the doe is drinking, indicating that they are thirsty and unable to reach the nipple. Having nipples at a low level is unsatisfactory, for the adult cannot drink easily at such a level and may easily push the nipple with its body, thus causing excess loss. Unless two nipples are placed in the cage when the young are eighteen to twenty days of age, the watering nipple must be lowered to a level of 5in (12cm) or so, but no lower.

It is best to install nipples at the front of the cage so that servicing them is easier. At the same time it is desirable, unless a large bore tubing is used (and even then it is best), to keep the tubing out of reach of the rabbits, who can damage it. A daily inspection of the nipples will find those that are leaking and those which are blocked. Rapidly depressing the operating pin will sometimes clear the problem.

There are two basic methods of eliminating the freezing-up of the water system, although in general a rabbit house which allows frequent freezing of this system is not suitable for intensive rabbit production.

Housing and Equipment

The first method is to use external or internal heating tapes. The external tapes are not very satisfactory except perhaps for the main supply. The internal heating wires are available in various wattages per metre and in certain standard lengths. They have to be fed through the water lines and should always be connected to a thermostat – unless these wires are of an unduly high wattage, they are sometimes inefficient in preventing the nipples freezing up.

The second system is the heating of the circulating water. For this, it is necessary to ensure that the water in each line of nipples is returned to a low level tank (preferably situated under the breaker tank), in which the water is heated by an immersion heater and pumped up to the breaker tank. An overflow must be connected between the breaker tank and the heater tank and at the same time a thermostat is fitted to the heater tank water inlet. This system is satisfactory if the pipes to which the nipples are fitted are of the appropriate size – unless the run is very short even ⅜in (1cm) diameter pipe might not be sufficiently large. Using this method, a 1500W (1.5kW) heater, with thermostat operating, will be able to keep the water for a fifty doe unit warm enough to prevent it freezing up.

Feed Hoppers

Although a wide variety of containers is used for rabbit feeding in commercial rabbit husbandry, only one type is suitable – the feed hopper. There are a number of different designs of feed hoppers; however, almost all have a 'J'-shaped body, which may fit on to the outside of the cage and project inwards, or which may remain at the face of the cage. The latter design is preferable.

The typical hopper may be 6–8in (15–20cm) wide and may contain 4.5–6.6lb (2–3kg) of pellets if completely filled. (There are also much smaller sizes for individual hand feeding.) It must be securely fixed to the cage front to prevent the animals from dislodging it, but at the same time it should be easily removable for regular cleaning. This is particularly necessary if, as sometimes happens, it is sprayed by the animals or becomes wet from a leak in the watering system.

The base of the hopper often has a fine mesh through which the dust from the feeding pellets may fall. If this is completely open, then all one finds is feeding pellet dust on the floor. It is much better if a hinged flap at the bottom of the hopper is used to contain the dust and emptied from time to time.

A good deal of the dust may be produced by the rabbits scratching at the food – this activity also scrapes some feeding pellets on to the mesh floor, through which they fall. This scratching, and the habit that some

Housing and Equipment

Young meat animals in a colony pen, with extra large circular feeders.

Housing and Equipment

young rabbits have of climbing onto the hopper, can to some extent be prevented by dividers running from the back to the front of the container and upwards. A further development of this is a tunnel attachment into which the animal has to put its head in order to feed. This, however, is only suitable for adult rabbits.

Feed hoppers are now almost invariably manufactured from galvanised iron. They should preferably be galvanised after manufacture rather than made from cut sheet. It is important that they should be well and strongly constructed and particularly that there are no sharp projections which can damage stock.

The Nest Box

The health of the newly-born rabbits depends to a considerable extent on the design and effectiveness of the nest box. On many rabbit farms two thirds of deaths in the first month of life occur in the first few days. Factors which are important in the design and construction of the nest box are as follows:

1. Its size must be such that the doe can adequately kindle and nurse her young.
2. It must allow easy examination and operation (such as for fostering, etc.) by the farmer.
3. It must ensure that the temperature can be easily maintained at about 86°F (30°C) in the centre of the nest – the young must be able to huddle together in colder weather, and to disperse in order to keep cool in warmer conditions.
4. The construction must be such that the nest box must be able to be thoroughly cleaned and disinfected.
5. Urine must be allowed to escape.
6. The box must be of such a form and in such a position that the young are prevented from getting out, but can easily get back in should they be dragged out inadvertently by the doe. When they are first venturing out the construction must be such that they do not hurt themselves. In some cases, with a slippery floor, the young can dislocate their hips.

There are a number of different designs of nest boxes, ranging from the very simple open-top box put into the cage to sunken nest boxes with lids for internal use. There is little doubt, however, that, although it may be more expensive than others, the external nest box is the one to be recommended.

A satisfactory external nest box is manufactured from galvanised iron

Housing and Equipment

A well-filled sunken nest with a litter of twelve.

Housing and Equipment

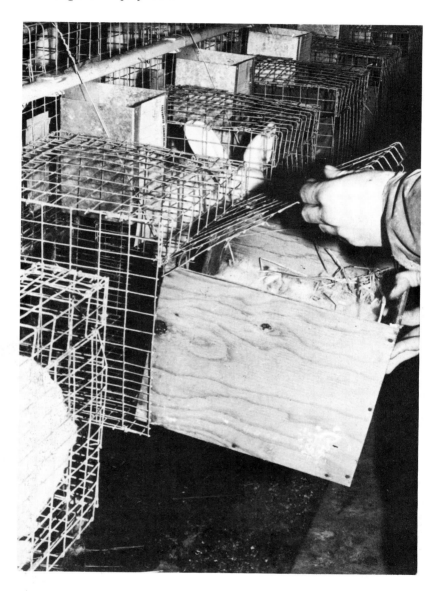

Another form of nest box which is withdrawn from the front for examination.

Housing and Equipment

and should be designed so that the base can be easily removed for thorough cleaning. An insert of wood in the base gives a good deal of insulation, as does an internal surround of wood on three sides. There should be a hinged lid to enable the breeder to examine the young, and preferably the lip of the entrance hole should be fitted as flush with the floor as possible. Appropriate drainage from the base can be obtained if a removable piece of galvanised iron is slotted at the edges, and is held in the box by flanges for use as the floor.

Occasionally, in hot weather, this type of nest box may become too hot – replacing the solid lid previously used with a simple mesh cover will alleviate this problem. This new cover must, of course, be firmly fixed to the nest box.

A plastic nest box with litter of eight New Zealand Whites.

Housing and Equipment

Fairly standard old-type nest box, showing the damage that can occur within a relatively short period.

The size of the nest box for the typical New Zealand White type doe should be about 7in (18cm) and the lower level, which should be flush 12in (25–30cm) high. The entrance hole, which is in the largest wall, should be about 7in (18cm) and the lower level, which should be flush with the cage floor, should be 2.5–3in (6–7cm) from the floor of the nest box itself. In this way the young rabbits are prevented from getting out until they are large enough.

Cleaning and the Environment

The last matter to be touched upon is the cleaning of the equipment and the environment in which the animals live. Constant hygiene and the routine maintenance of the equipment, and its repair as soon as damage occurs, are the only possible rules in intensive livestock husbandry. There is no doubt that all items of equipment, including cages, are best cleaned in a place where special facilities are available. Plenty of water, high-pressure hoses, and containers in which suitable disinfectants can be held to soak equipment are all essential.

Housing and Equipment

Calcium and other deposits from urine do accumulate and if left become almost impossible to remove. Soiling of mesh and equipment by manure is also extremely difficult to remove when it becomes very dry, and the only answer is to allow it to soak thoroughly before cleaning.

The removal of fur by flame, whilst satisfactory for that purpose, should not leave the breeder under the false impression that the cages have been sterilised by the heat. To attempt to sterilise a cage in this way would almost certainly ruin the galvanisation.

The procedure for sterilisation is as follows: all items must first be cleaned thoroughly (hot detergents are necessary for this). The equipment must then be sterilised, a process which is followed by thorough rinsing and then thorough drying. The drying, a matter of days rather than hours, must be undertaken in a place where the items will not be reinfected. This may seem a long-winded process, but if it is carefully thought out and carried out in a methodical fashion, it need not be as laborious as it sounds.

The use of flame and extended wire brush to clean out the cages.

Housing and Equipment

One point should be made about wire brushes – they can and do damage galvanised metal. A rusted piece of mesh or steel is harmful to the animals.

Apart from the cleaning of the cages and equipment there is the cleaning of the building, and the air generally, to do. This includes the elimination of bacteria and insects, which tend to build up to insupportable levels, and of the dust and the fumes that arise.

Two kinds of sprays are used for this purpose. The first are the bactericidal sprays, and the second the insecticidal. There is little to be said, other than that they should both be used strictly according to the manufacturers' instructions, with a frequency which is appropriate to the season of the year (in the case of the insecticidal) or in accordance with the recommendations (for the bactericidal sprays). Whenever an opportunity arises for the rabbit house to be totally fumigated, that is, when it is empty of all animals, the task should be done. As it is an unpleasant process, care should be taken to close the house thoroughly and then to leave it for a sufficient time to allow all indications of fumigation to disappear.

Dust and fluff will always arise. A thorough removal should be undertaken whenever possible. One of the great wasters of light is the accumulation of dust on lights.

Finally, there is the question of the elimination of vermin from the house. These cause loss and damage not only by their eating the available food, but also by the disease they spread or, in the case of rats, by the damage they may do to young rabbits. The contamination of food by the droppings of vermin is a certain cause of loss of stock. All steps should be taken to prevent the entrance of vermin to the house and to attack and destroy them at all times.

2 Breeds and Breeding

BREEDS OF RABBIT

There are well over forty different pure breeds and something over 200 different colour varieties of those breeds; this is not including the Rex, of which there are twenty-six colour varieties in Britain today. A number of these different breeds have been used at one time or another for commercial rabbit keeping, but for meat production at the present time, only two breeds form the basis of the vast majority of farms. These are the New Zealand White and the Californian.

Much of the stock used in commercial rabbit farming should not actually be called New Zealand White or Californian. Those names should be reserved for animals bred to a standard of perfection, largely for their external characteristics, and should not be used for animals that are bred for their utility characteristics. Many of the strains of rabbits

New Zealand White – the most popular commercial breed world-wide.

Breeds and Breeding

used in some of the best farms have had several other breeds of rabbit introduced into them and are certainly cross-bred.

The characteristic required of commercial meat-producing stock is the ability to produce the maximum quantity of young per animal – those young must have good meat qualities and use food as efficiently as possible. This broad statement does imply that a good capacity for survival is one of the most important characteristics.

The New Zealand White

Adult bucks of the New Zealand White type weigh 9 – 10lb (4–5kg) and the does a pound (half a kilo) or so more. The breed produces an excellent carcass when properly grown and, as its name implies, it is completely white.

The original New Zealand White was first produced in America for the farming of young meat rabbits, particularly in the west of the country. Its first importation into this country was in the late 1940s and from here and from America the breed has spread to many parts of the world.

Californian – the second most popular commercial breed in this country, it is renowned for its mothering qualities.

Breeds and Breeding

The Californian

The Californian is slightly smaller, the largest weighing, at most, 10.5lb (4.75kg) – quite often good bucks weigh as little as 7.75lb (3.5kg). In general, this breed produces a superior carcass. Again it is a mostly white rabbit, but its points (its ears, nose, feet and tail) are coloured.

The Californian was first developed, again in America, in the 1920s, by the use of New Zealand Whites, Himalayans and Chinchillas. It was brought over to this country in the late 1950s, at that time solely for its commercial possibilities. The carcass conformation of the average Californians is probably better than that of the average New Zealand Whites, since it is more blocky and meaty, with a greater width across the loins. The New Zealand White is, however, more widely used.

French Lop – imported in the very early 1960s to give greater width to the carcass.

Breeds and Breeding

Other Breeds

A number of other breeds have been used to bring variety to the commercial rabbit. The French Lop, for example, has, in some strains, improved the conformation of the carcass greatly with its peculiarly 'cubic' body.

There have been a number of attempts in this country to produce specialised lines of rabbits for commercial meat production. Most of these have ceased to exist. Some of these lines or 'breeds' are or were excellent, but others allegedly having greatly improved performance are offered for sale to unsuspecting newcomers and this results only in profit to the originator. The situation is likely to change, as it has done on the Continent where great attempts have been made, in some cases with considerable success, to produce animals which give greatly improved performance.

For specialised purposes other breeds are used. In the laboratory animals market, a smaller breed weighing about 4.5lb (2kg) is generally

Dutch Rabbit – a small breed, having excellent carcass qualities and mothering characteristics, and much used for the production of pure lines for crossing.

Breeds and Breeding

preferred, and for high quality furs the Rex breed or some of the normal fur breeds are used.

But for commercial rabbit farming it is desirable to have a population of animals all with those superior characteristics (none of which are typical 'breed' characteristics) which will lead to maximum profit.

REPLACEMENT AND ADDITIONAL BREEDING STOCK

The most successful rabbit farmers practise a continuous and fairly severe culling of breeding stock. Does and bucks which do not perform to a high level must be disposed of, as soon as the fact is recognised, in order to make room for more useful stock.

On average, it is unlikely that the best farms will have a life production of more than nine or so litters per doe. In some cases the number may be smaller. This means that some two per cent or more of the output of all does is required for replacement breeding stock. To allow the farm to run efficiently, it is necessary to breed the replacement stock on a continuous basis and to replace the existing breeding animals which are to be culled with the developing breeding stock.

At the commencement of a unit, if no animals are to be brought in from outside as replacements, it is necessary to select animals for retention as future breeding stock. This choice is based solely on the very early performance of the doe and young and the appearance of the young animals produced. As the unit matures, this situation can change gradually to one in which a full performance recording and selection programme can be instituted.

INHERITANCE AND HERITABILITY

The subject of herd improvement and the inheritance of commercially desirable traits in animals is wide-ranging – those readers who wish to have a detailed understanding of the subject *see* Further Reading.

All the characteristics and qualities of an animal are produced by the interaction of its inheritance and its environment. All those factors which influence it in any way after it is born are environmental and the animal is affected by such things as nutrition, housing, management and contact with disease.

The inheritance of any characteristic may be controlled by one gene or by many. Colour is usually controlled by one gene only – for example, there is a specific gene for black pigmentation and the recessive

Breeds and Breeding

counterpart of this gene produces a brown colour. A dominant gene suppresses the action of its recessive counterpart – therefore, a rabbit receiving the black gene from either its dam or sire will be black. However, it does need to receive the recessive brown gene from *both* parents before it can be brown.

Most of the characteristics of the rabbit that are useful to the commercial rabbit farmer are controlled by a number of different genes (if, indeed, there is any inheritance at all). Inheriting these important traits is not, then, a simple either/or as it is with the inheritance of black or brown.

Heritability is the degree of likelihood that a certain characteristic will be influenced through inheritance from the parents. It can be expressed as a percentage, on a scale of from zero to one, or it can be described simply as low, medium or high. Improvement in a herd can only successfully be obtained by improving the environment in all its aspects so that the full inherited potential of the animals can be fulfilled, and then selecting future breeding stock on the basis of the performance of animals in respect of those characteristics which have the highest heritability.

The measurement of heritability is a complicated matter and estimates tend to vary a good deal. However, the actual degree of heritability of

	Heritability		
	Low	Medium	High
Conception rate		x	
Weaning weight		x	
Individual weight at slaughter		x	
Individual early daily growth rate			x
Litter size at birth	x		
Number alive at 21 days and slaughter	x		
Number of teats			x
Temperament of doe (maternal characteristics)		x	
Food conversion rate		x	
Dress out percentage			x
Carcass conformation			x

Commercially valuable characteristics of the rabbit and their heritability.

any characteristic (if such a degree were possible to measure) does not really concern the rabbit farmer. He is interested simply in whether the heritability is low, medium or high. The table opposite gives estimates of the heritability of certain characteristics in the rabbit.

IMPROVEMENT OF THE HERD

The commercial rabbit farmer has several options in terms of breeding and improvement of his herd. The first possibility is simply to mate any doe to any buck in a random fashion, after discarding only those few animals which have an unsatisfactory appearance and those which he remembers to be poor performers. There may be a slight improvement in the herd because the breeder will perhaps make fewer mistakes as time passes. However, he will probably not be able to continue for long, as the enterprise is unlikely to meet with success using this method.

His second option, and the method most commonly adopted, is to select the most suitable animals, both male and female, cull the breeding stock more severely, keep some records and mate the animals in some sort of systematic fashion. Improvement will occur at a slow rate over the generations but will be variable – progress is likely to be quicker if he adopts the third method. This is a completely systematic approach, analysing the records of his stock so that only rabbits of the best quality are used to breed replacement animals.

The rabbit farmer is fortunate in that he has only a small amount of time to wait between generations before he gets some initial idea as to the breeding value of any animal. He is also fortunate in that the number of young per adult is large. No judgement should ever be made of the doe on her first litter – the second litter may be better, but on average it will still not be quite as good as later litters. It is, however, reasonable to use the second and third as a basis for a preliminary selection. Thus, by the time the doe is nine months old the breeder may not have enough information on which to base a firm selection, but he will at least have an idea as to her possible potential.

Thereafter, at intervals of six to eight weeks (possibly with rest periods every three or four litters), the doe will be weaned of another litter – the records of her performance will increase and will provide more detailed information on her comparative breeding value.

In the same way, by the time a buck is eight or nine months old, there can be a sufficient number of breeding records of him (say, sixty or more) to enable an initial decision to be made as to his value as a future parent of breeding stock.

Breeds and Breeding

TESTING

Performance testing and progeny testing have both been used with success in the establishment of improved lines, although in many cases these lines have unfortunately been lost or ruined by the wrong introduction of other animals. Performance testing is actually carried out by the rabbit breeder when he counts and records the number of young born (alive and dead) and does the same for weaned litter weight – it is the assessment of the breeding value of an animal on the basis of its own performance. It can only be used successfully to improve the herd when the attributes which the breeder wishes to select have a relatively high heritability.

Progeny testing differs from performance testing in that the performance of the progeny is used to assess the breeding value of a parent (usually the male). It is the only possible test system for a male, for such characteristics as teat number, milk yield, weight of the young at twenty-one days, and so on, can only apply to does. It is also useful for selecting animals for carcass dress out percentage (the weight of saleable carcass in proportion to the liveweight) and conformation. Although a number of national schemes use progeny testing rather than performance testing (and have had a great deal of success with it), there are difficulties. A large number of tests have to be completed and the amount of work involved can be extensive. However, this is not the case with performance testing and the commercial rabbit breeder is advised to make the latter the avenue through which he will proceed.

The subject of records and their analysis is dealt with later, but it must be on the basis of these that selections for the production of future breeding stock are based. There are two ways of selecting the parents of future stock. The first is by simple comparison of the final sets of figures produced for does or bucks, and the second is by the reduction of each set of figures to a single index. The subject of the preparation of such an index is outside the scope of this book – in any event, it would be somewhat unusual for anyone except the most advanced breeder to use such an aid.

Any animals likely to pass on such disorders as malocclusion, and those whose disease record is other than perfect (for example, those with 'sore hocks'), having been eliminated, bucks and does for further breeding are selected according to the records. The procedure is then to allocate bucks to the different does to follow whichever breeding system has been chosen.

Breeds and Breeding

SYSTEMS OF BREEDING

Systems of breeding fall easily into separate categories. The first is *inbreeding*, in which animals more closely related to each other than average, such as brother and sister (the closest form), or parent and offspring, are mated together. *Line-breeding*, which is mating progeny always back to the same particular animal (usually a buck, or his sons when he is no longer available), is a form of inbreeding. Both forms throw up any defects which happen to be present in the stock and 'inbreeding depression' may arise, involving a gradual reduction in the animals of such qualities as fertility and resistance to disease. Whether this problem appears or not will depend to some extent on the degree of inbreeding. Some degree of inbreeding is certainly necessary if any level of genetic purity is to be achieved but for commercial rabbit breeders (except those with considerable resources and skill) the disadvantages of inbreeding probably outweigh the advantages.

The other categories of breeding are *assortive breeding*, or the mating of animals which, although they are not more closely related than average, look alike; *corrective breeding*, in which unrelated animals that each have different good points and bad points are mated in the hope that each will contribute a compensatory effect; and *cross-breeding*.

Cross-breeding implies the mating of animals from two quite separate groups, either pure breeds or pure strains. The underlying theory is that if two animals of widely differing genetical constitution are mated, the resulting progeny show 'heterosis' (or hybrid vigour), which tends to produce a result that is better than the average of the two parents. It is true that this most usually happens, but on occasions the reverse can occur.

A question arises with cross-breeding as to what is the next step to be taken when the first generation (called the F1 generation) is itself to be used for breeding. In the case of rabbits only the females would be used. The options are to breed back to a buck of either of the lines or breeds (*back-crossing*), to breed the F1 generation together (*inter-breeding*) or to bring in completely new and different lines in each generation (*rotational cross-breeding*).

There is one further activity which, although not a system as such, should be considered. This is *out-crossing*, which occurs when a new buck from an outside source is introduced into a unit. The result is often similar to that of cross-breeding in that an F1 generation, with increased uniformity, is produced.

These different systems each have their own adherents, although in some cases combinations of the various systems are used. The most

Breeds and Breeding

common method in the unit that is producing its own breeding stock replacements (or indeed animals for expansion), is assortive breeding. In any system rigorous culling is the key to fast improvement.

3 Nutrition and Feeding

One of the most important factors for success in commercial rabbit farming must certainly be the correct nutrition and feeding of the animals. Food is the most important factor in the maintenance of good health and good production, and the costs of feeding in most systems represent at least seventy per cent of all expenses. It is essential, therefore, for the breeder to pay the greatest attention to this matter if he is to enjoy success.

The great majority of commercial rabbits are fed on complete pelleted feeding stuffs and this has been the case for over twenty years. There are a number of manufacturers of pelleted feeding stuffs and it is very unusual for trouble to arise from the use of such foods (provided they are stored correctly and used reasonably fresh). Problems do occasionally occur, however. What appears to suit some farms does not, for one reason or another (usually unknown), suit others. It is necessary, therefore, to keep a watchful eye on supplies.

FOOD CONSTITUENTS

All food stuffs for rabbits have protein, fats (or oils), fibre and carbohydrate as the major constituents, together with minerals and vitamins and often some additives such as coccidiostats (*see* page 54) and growth promoters. All foods also contain varying amounts of moisture. In pelleted feeding stuffs, the moisture content is usually around eleven or twelve per cent, but may be as much as fourteen per cent or so. It is important to know this percentage for any calculations which will have to be made – for example, metabolisable energy is given on the basis of the dry matter and not of the weight of the food as fed.

Protein

Protein consists of varying amounts of amino acids (of which there are around twenty-one), eleven of which are essential to the rabbit. There is of course both animal and plant protein, and a mixture of both is usually incorporated in pelleted feeding stuffs.

Nutrition and Feeding

Fats and Carbohydrates

These supply the energy of the ration although any surplus protein may also be used in this way. Fat supplies something over twice as much energy per unit as carbohydrate does.

Fibre

This must be considered as being in two forms – digestible and indigestible. Of the total fibre content of the ration, at least eighty-five per cent needs to be indigestible in order for the digestive processes of the rabbit to function satisfactorily. Some fibre (for example, maize and sugar beet pulp) has a very high digestibility, whilst the fibre that occurs in other foods such as oats or straw, is very indigestible. It will be seen from the table on page 52 that the minimum recommended levels of indigestible fibre are those in the ration for lactating does with young, an amount of ten per cent. The indigestible fibre content should never be reduced below this level in any ration.

Minerals

These are required in quite large amounts whilst others are required in trace quantities only. Apart from a satisfactory level of minerals of the different types, it is essential that these should be more or less in the correct proportions. An inbalance can cause as many problems as a deficiency.

Vitamins

Vitamins are also required and although natural foods contain differing amounts of all those which are needed, supplementation of some vitamins has proved of benefit.

Most feeding stuffs manufacturers have access to a great deal of scientific information relating to the needs of the rabbit and it can reasonably be assumed that feeding pellets sold by a reputable manufacturer will be adequate in their composition.

MEASURING FOOD VALUES

There are two systems used for the measurement of food values. The first, and oldest, involves calculating the 'total digestible nutrients' of a

Nutrition and Feeding

ration, which gives a value per kilogram of food. The different digestible components of the food are multiplied by the amount of units of energy they give, and added together. One kilogram of TDN will yield approximately 4,000 kilocalories of metabolisable energy. The food value is expressed in TDN/kg and in the percentages of protein, fibre, minerals, and so on, that are present.

The second system, which has certain advantages, expresses the total energy available for use by the animal in terms of 'metabolisable energy' or ME. The unit of measurement is the megajoule (MJ); since it is the amount required per unit of time that is needed, the energy requirements per day are given (MJ/day).

Therefore, given the ME of each food, together with its digestible protein content, comparisons between foodstuffs can easily be made with regard to the prices, and the food requirements of animals can be matched against the declared food value. The relationship between TDN and the MJ is as follows: 1g of the TDN of a food, multiplied by four, yields 1kcal of metabolisable energy, and 239kcal equal 1MJ.

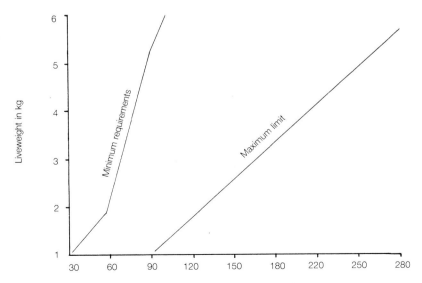

Minimum and maximum amount of dry matter required or permissible in rations per day in relation to liveweight.

Nutrition and Feeding

FOOD REQUIREMENTS

Food is required for maintenance, growth, lactation and so on. These activities require different quantities of the various food constituents. Thus the lactating doe and rapidly-growing young need far more protein than, for example, a resting adult. In the same way, the lactating doe has a much higher requirement for minerals than the adult buck. All animals need a certain amount of bulk to satisfy appetite, but there is also a limit to the amount that they can physically consume. The food given must therefore be sufficiently rich in protein and energy to satisfy the requirements of the animal, but not exceed the bulk that it can eat. The greatest difficulty in this respect concerns the lactating doe who is also pregnant, and especially towards the third quarter of the pregnancy if this coincides with the peak of lactation. The graph on page 51 gives details of the minimum and maximum amounts of bulk which on average are acceptable to rabbits.

The table below gives the desirable proportions of the different constituents in the rations recommended for each class of animal – these will most nearly satisfy their requirements for maximum production, whatever form that might take, and should keep them in good health.

Unfortunately, it is not usually possible to have a series of rations for different groups of animals.

	Crude protein	Total crude fibre	CF non-dig.	Fat	Minerals	Metab. energy (MJ) per kg[1]
1 Lactating does and young	18	12	10	3	8	10.5
2 Does in kindle not lactating	16	14	12	3	7	10.0
3 Weaners to slaughter age	16	14	11	3	6	10.0
4 Adult working bucks	14	14	11	3	5	9.5
5 Resting adult stock	13	15	13	3	5	9.0
6 Growing stock for breeding at 10 weeks plus	15	14	12	3	6	9.5
7 One universal food[2]	16	14	11	3	7	10.0

[1] Dry matter basis
[2] Not entirely suitable *by itself* for resting adults and late stage youngsters for breeding

Ideal proportions of constituents of food for different classes of stock.

Nutrition and Feeding

THE COSTS OF FEEDING

Whilst the energy and protein content of a food is a major consideration in relation to the price of that food, the cost must be measured in terms of the total outlay involved in producing a unit gain in liveweight. Very cheap foods (measured solely in pence per kilogram of food) give a different result when measured in terms of pence per kilogram of liveweight gain. Furthermore, the comparison of two feeding pellets in costs per kilogram of liveweight gain is meaningless unless the comparison is carried out under the same conditions. Food conversion ratios depend upon a number of other factors as well as upon the quality of the food.

Certain alternative foods when costed in pence per kilogram can appear to be very cheap, but the small amount of money saved in producing liveweight gain may easily be more than cancelled out by the cost incurred by the increased labour involved.

Food Prices

It will be found that the price of feeding pellets varies from one supplier to another. It is wise, therefore, to investigate prices from a number of different sources, although it is never advisable to change the particular type of pellet being used simply on the basis of price. The variations, however, may be as much as ten or fifteen per cent and sometimes even slightly more. (It must be said that some of these variations are also due to differences in quality.)

Food Conversion Ratios

The food conversion ratio (FCR) is an important matter for the rabbit breeder. It is the ratio of the amount of food required to produce a unit amount of liveweight gain. In practice the total amount of food eaten by the doe and litter up to slaughter age, plus an allowance for the buck, is placed against the total weight of the young rabbits sent for slaughter. Claims have been made that an amount of food of under three kilograms has been used to obtain each kilogram of liveweight gain. This is an exceptional figure and when it is achieved by a large group of breeding does there is always the suspicion that the measurements have not been strictly accurate. Three and a half to one (food to liveweight gain) is, however, obtained on a reasonably large scale by the best of farmers.

Nutrition and Feeding

FEEDING PELLETS

Specialised Pellets

The table on page 52 shows that feeding pellets to be used for the various categories of stock should have differing compositions for the best possible results. However, this is not usually a practicable proposition, for the costs of the operation, except for the very largest farms, might possibly outweigh the financial gains. As it is there is almost always the need for farmers to have two types of feeding pellet – one with a coccidiostat and one for use during the withdrawal period.

The suggestion is often made, and it is certainly not unreasonable, that there should be 200 to 250 does to justify each additional type of pellet put into use. It is not economic to purchase two specialised pellets unless the unit has some 400 breeding does, and it would not be of value to bring in a third pellet until a number of 600 or 700 is reached.

One of the main problems of using a high-energy pellet throughout the unit is the tendency of dry does (of which there should be few) and young stock coming up to breeding age to become over-fat. Excess fatness is a very unsatisfactory condition for breeding stock and, indeed, leads to reduced conception rate and even sterility in bad cases.

Food Additives

In recent years, it has become the practice to add certain additives to pelleted feed mixes. These are strictly controlled by law. An additive may be either a veterinary drug to control disease or a growth promoter. In the case of coccidiostats (drugs that are used to control coccidiosis), two main ones are used – clopidol and robenidine. These drugs are included at levels which prevent the development of the disease and all bags containing feed with a coccidiostat have the letters ACS printed upon them. In each case the farmer must allow a period before the animals are sent for slaughter during which the drug can be totally eliminated. A statement of the withdrawal period necessary – which tends to be seven or ten days – is required by law to be incorporated on the feed bag label.

Physical Characteristics

Feeding pellets can be of varying sizes but the ideal is about 0.25–0.3in (6 or 7mm) long by about 0.15in (4mm) in diameter. Rabbits tend to dislike dusty pellets, and if the pellets are allowed to crumble and dust some food will be wasted. The apearance and dustiness of pellets is some

Nutrition and Feeding

indication of their quality, particularly in those cases where food bins are used and the pellets delivered in bulk. Dust from food is not only wasteful but is harmful to the rabbits themselves.

WATER REQUIREMENTS

It is rather surprising to consider that at one time many (if not all) rabbit breeders thought that not only did rabbits not need water but that on most occasions it could be positively harmful.

Ample supplies of fresh clean water are of the greatest importance. The requirement for resting animals in average temperatures varies, but approximates towards three times the total dry matter content of the food. With an increase in temperature there are naturally considerably greater water requirements, as there are also during lactation. The milking doe with a good litter may easily be drinking as much as seven pints (four litres) a day as weaning approaches.

Usually, very little attention is paid to the question of water temperature. In very hot weather, when the requirement for water is at its highest, consumption by the stock is reduced rather than increased if the temperature of the water becomes too warm. In such cases some cooling, even if it only involves allowing mains water to pass through the system, has proved effective. The opposite applies to the drinking water in winter, when it is advantageous to have the temperature a little warmer than just above freezing point.

ALTERNATIVE FEEDING SYSTEMS

So far, mention has only been made of pelleted feeding stuffs. Whilst the majority of rabbits are fed almost entirely on these some alternative foods are used by a few farmers. The main incentive for a change to alternative foods is the lower cost. Any reduction in the cost of food per pound of liveweight gain leads to an increase in profit, provided of course that any extra expense of using such feeding stuffs is not greater than the saving in food cost.

The traditional foods of rabbit keeping have largely been replaced by pelleted feeding stuffs but green forages, roots, kales, silage and hays are all still used by some people. These natural foods are usually only fed to replace a portion of the pelleted foods.

There is no doubt that under really skilled and knowledgeable management some success may be achieved with these alternative foods. In general, however, they are only for the most experienced farmers or

Nutrition and Feeding

for those producing on the smallest scale. The use of by-products and waste material from a general mixed smallholding can be considered in the same way.

STORAGE OF FOOD

The manufacturing process of feeding pellets is such that it is most unlikely that any food supplied fresh and in closed bags, by a reputable manufacturer or merchant, will be contaminated. This always supposes that the bags of food have not been soaked at any time for this might cause damage. Any bags that are broken open or holed during delivery should be stored in bin containers and used first.

All food should be stored in a secure, dry and vermin-free place and although food has a fairly long life (it would not be unreasonable to expect it to last for six weeks), it should be stacked and used in the order in which it is delivered.

Bulk feeding silos may be used on some farms. A farm has to be fairly large even to think about the use of bulk deliveries of feeding stuffs. The financial inducement to take food in bulk is a discount on the bag price of anything from five to twelve per cent. There are, however, disadvantages, and the cash value of losses incurred through wastage may certainly exceed the lower discount, if not the higher. Then there is the cost of the silo in which the material must be stored. It is also doubtful whether there is any saving of labour – indeed it is often suggested that the reverse is true. It is clear, then, that the savings of bulk buying are slight and it takes a long time to repay the initial expense.

If bins are to be used it is important that they should be adequately cleaned between deliveries. During cleaning, steps should be taken to ensure that no stale food is left inside to become mixed with the fresh food. The cleaning of bulk bins needs more attention than it customarily receives. Unless a bin is divided so that one half can be completely emptied and cleaned thoroughly before a delivery, then obviously either a mixed system (some food bought in bulk and some in individual bags) must be used, or there should be more than one bin.

It is very important to ensure the continuity of supply of food to the rabbit unit. In very approximate terms, an intensively-run unit may use around 22lb (10kg) of feeding pellets per breeding doe (and followers) per week. Adequate storage facilities of the right kind must therefore be available, and must be able to hold the quantity to be received at each delivery with some space left over.

It is rare that a floor is sufficiently dry to stack food directly upon it. It

Nutrition and Feeding

is far better to use some sort of platform, standing slightly above the floor surface – wooden pallets are ideal for this purpose.

FEEDING PRACTICE

Commercial rabbits should invariably be fed from a well-designed feed hopper, even though they may be on a restricted ration and therefore fed a daily amount.

The aim of good feeding is to produce a particular result – either the greatest production or the right condition. The different classes of stock require different quantities of the nutrients. In some cases, it is necessary to regulate feeding to prevent the rabbits getting too fat, but it is almost impossible to overfeed young weaned stock destined for early slaughter, or a heavily-producing doe. It is very easy to overfeed animals from twelve weeks of age up to their first breeding and excessive fatness leads to lowered conception rates and even sterility in some cases.

Some stock, therefore, are fed ad lib – that is to say, they have food in front of them all the time – while others are fed a restricted daily amount. As a generalisation, the following quantities of feeding stuffs for different classes of animal are appropriate:

	g	oz
Lactating does with young	ad lib	
In kindle does, without litters, first 15 days	150	5.3
In kindle does, without litters, second 15 days	220	7.75
Fattening stock for market	ad lib	
Growing stock from 10 weeks of age	115	4
Working bucks	125	4.5
Non-working adults	115	4

Whilst the above figures give the approximate amounts necessary it must always be remembered that individual animals can require varying quantities. The stockman must keep a close watch on the feed hoppers into which hand-fed amounts are placed and adjust the amounts if necessary.

Good feeding practice is most important. Although the rabbit is an animal that naturally feeds on numerous occasions throughout the day, those that are hand-fed daily – that is to say, those on fixed amounts – do learn a routine, and its disturbance does nothing to help the best animals. The same routine should, therefore, always be adopted and it is as well here to remind the rabbit farmer that cleanliness should be an important part of it.

Nutrition and Feeding

Any container used to ration out feeding stuffs should be checked carefully to ensure that when it is levelled accurately to a mark (and undoubtedly the best way is to level it to the rim) it contains the appropriate amount of pellets required. It is surprising that some pellet-measuring devices in use are inaccurate to as much as fifty per cent. The weight of food given can also vary in volume, therefore different batches of feed should be checked from time to time.

Animals vary greatly in their tendency to run to fat and a good stockman will modify the rations for those which require it as soon as there is any indication of this becoming a problem.

Mature stud bucks present a slight problem in that their habit of spraying urine sometimes dampens and clogs food in hoppers. As a result, these may require more frequent cleaning, and it may be an advantage to increase the height of the hoppers from the floor.

The final point concerns the change of foods. Troubles will arise if this is too abrupt – however, any problems will not be caused so much by the change of food (as in, for example, a simple adoption of different pellets), but rather they are more likely to be the result of any variation in the type or the nature of the food. It is a wise precaution to introduce change over a period, incorporating an increasing proportion of the new food every few days over a period of ten days or a fortnight.

4 Management and Stockmanship

Management consists of all those activities, both physical and mental, which are necessary for the successful running of a rabbit unit. Stockmanship is that ability in some people, which enables them to breed and rear animals successfully. It is the same characteristic, expressed only in a different field, as 'green fingers' in a gardener. Stockmanship has a great deal to do with observation and the interpretation of what is observed. The management of rabbits must be based upon good stockmanship. Ninety per cent of the success of a rabbit enterprise probably depends upon good management and stockmanship. Without one or other, the enterprise will fail.

Anyone who wishes to succeed in rabbit farming must develop in every way possible their stockmanship and abilities of observation. Attempts must be made to observe the animals at all times in a systematic fashion. If a system of general observation (and the examination of individual animals) is developed, there is far less chance that an important point will be overlooked.

Stockmanship is also based to a large extent on habit, which must, of course, be good. For example, the good stockman will never handle animals immediately after handling diseased stock without thorough washing. Diseased animals will never be left as a source of infection. The failure or malfunction of equipment such as watering systems or ventilating systems is immediately noticed and rectified. Signs of ill health (lack of appetite, for example), are recognised and early action taken. Such observation and action must become quite automatic.

The other side of management, the planning side, is no less important. Attention to detail, as in the observation of physical condition, is essential to success. The more one learns of the techniques of rabbit husbandry and about the animal, and the more one analyses the causes of the failures that will certainly occur, the more successful one will become.

HANDLING

There are a number of methods of holding rabbits and the best way varies depending on the age and size of the animal and the purpose for

Management and Stockmanship

which it is held. In all cases, the actions should not be overhurried – firmness and gentleness combined will make the rabbit feel secure and as a result it will not struggle. A rabbit which is frightened and struggling can hurt both itself and the operator and often the best thing to do in such a case is to put it down and then pick it up again more securely. Rabbits can be lifted as follows:

1. Hold the ears, with the thumb at the base of the ears in front and the rest of the hand firmly 'cupping' the back of the head. In this case the weight must always be taken either by having the other hand under the hindquarters, or by grasping the animal across the rump. This method is suitable for all animals from about four or five weeks of age upwards and for all but the smallest breeds.
2. Grasp the loose skin over the shoulders. Although on the Continent this is often the only hold used, even on the very large animals, in general it is better to support the weight with the other hand or arm under the hindquarters. When handling rabbits in this fashion it is necessary to ensure that the skin is grasped with the flat of the fingers and not with the tips, otherwise the skin may be damaged by the finger nails.
3. Grasp the rabbit with one hand across the loins. This is a useful

Lifting, ears and hindquarters.

Management and Stockmanship

Lifting, skin on back.

Lifting, loins.

Management and Stockmanship

Carrying, under arm.

method in suitable cases but can be a dangerous practice if the handler is rough, and the internal organs and the muscles of the animal may be damaged. It is also not suitable for an animal over a weight of about 5.5lb (2.5kg) or any doe in kindle.

4. Lift the whole animal by grasping it round the shoulders and chest or the whole of the body with the back of the palm of the hand. This technique is used for small animals and again it is most important that the animal should be held gently but firmly.

Once the animal has been lifted, it should be carried against the operator's chest. If it is also held against the left arm, it will feel and be more secure.

When the animal has to be examined or placed on any bench or table, the surface should be covered with firmly secured sacking which will prevent the animal sliding about. A sacking-covered board can also be useful for this purpose when placed on a cage top. When the animal is lifted with the right hand grasping the ears with the thumb on the forehead, the animal can be quietened by stroking the forehead with the ball of the thumb.

Management and Stockmanship

PURCHASING STOCK

The purchasing of the stock with which to establish a rabbit farm is of the utmost importance, for the correct initial selection of the animals is fundamental to success.

It cannot be emphasised too often that, whilst there are some extremely reliable breeders who will treat newcomers with consideration, helpfulness and honesty, there are other establishments, some of them quite large and long-established, where the excellent salesmanship is only equalled by the very poor quality of the animals they sell.

It is essential for intending purchasers to visit a number of establishments before agreeing to buy – a great deal can be learnt from such visits. No matter which farms they go to, their future breeding stock should be purchased from an accredited breeder of the Commercial Rabbit Association. The Accredited Breeders Scheme was set up not only in an effort to improve the quality of breeding animals, but also to institute a fair trading system for all intending purchasers. The affiliated breeders agree to certain rules, and to arbitration by the CRA should any difficulties arise.

On his visits the prospective rabbit farmer should ask to see the records of the farm and should make an attempt to evaluate not only the management, but also the stock itself. He should look for any signs of ill health and for an appearance of well-being amongst the animals. A very high proportion of the does should have growing litters or well-filled nest boxes and a good proportion of all cages should be occupied.

Unless the breeder can produce good records it is unlikely that he can produce good breeding stock. The purchaser should have a very inquisitive mind at all times.

Future breeding stock are often purchased at eight weeks of age, but this is really too early. Twelve weeks is probably the best time to take youngsters, although sixteen weeks is not too late. It is not wise to buy fully adult animals.

One of the problems for the newcomer is to determine with any degree of certainty the age and health of the animals he receives. For this reason it is a sensible move for the beginner to obtain the independent assistance of some experienced person who can thoroughly examine all new arrivals.

SEXING

Young rabbits a day old can be sexed (in fact it is easier on the first day than a few days later), but it is a difficult task and rarely, if ever, needs to

Management and Stockmanship

be done by the commercial rabbit farmer. At a few weeks of age the job is much simpler. The sex organs are gently everted by using the finger and thumb and the round shape of the buck's penis will be distinguishable from the v-shaped female orifice. A little practice enables the handler to hold the animal (when it is small) on the palm of one hand whilst using the other to sex it. When larger animals are to be sexed single-handed then it is often best to lie the animal on its back on the operator's lap, or on a cage top, holding the animal steady with the left hand grasping the ears, the thumb in front and the palm cupping the back of the head.

MATING PRACTICE

The act of mating usually occupies a very short period. An important rule is that the buck should not be introduced into the cage in which the doe is housed. Some breeders have developed trolleys on which several cages are mounted (preferably with solid partitions between them). The bucks are placed in these and taken to the doe cages and the does are then introduced into the trolley cages. It is, however, rather better to have the bucks' cages interspersed between the cages of the does.

Double mating, that is, allowing a buck to serve the doe twice, or the

Handling for sexing.

Management and Stockmanship

use of two bucks on the same doe within the hour, is sometimes advocated, but this practice is wasteful and of no value.

Should a doe not stand to the buck almost right away, then it is probably best to remove her and try her again the next day.

Whilst the doe has no particular regular oestrus cycle, the eggs in her ovary mature and degenerate in a fairly consistent way. When she is highly in oestrus (with the vulva a pinkish, purplish colour) then she is almost invariably fertile and receptive to the buck. A pale, dry vulva tends to indicate the reverse. There is a variation throughout the year in the sexual response of the doe. To some extent, this seasonal variation can be reduced by uniform lighting and less variation in temperature than is natural. It has been suggested that variable light patterns may produce a more uniform response, but there is as yet no firm proof of the rules that need to be followed to produce this.

A young buck should not be used more than two or three times a week, and the first services should be well spaced out. More mature bucks can be given up to seven or, at the most, eight services a week and, indeed, when the need arises two and even three services a day are permissible with a strong, virile buck.

The ratio of does per buck is most often given as ten to one. There are, however, advantages (when very intensive breeding is carried out) in making the ratio less, say, seven or eight to one, with the reserve bucks always being given at least a service or two per week.

Matings are sometimes forced or assisted – the doe is held still, the hind quarters being slightly raised to assist the buck to mate. Some research workers have found that this does not increase the number of successful conceptions, whilst others have found it to be of assistance.

Conception rate (usually given as a percentage) is the number of does that become pregnant in relation to the number of does mated. In good establishments it should be about eighty-five per cent; in other words, about 118 matings are necessary for every 100 litters. The conception rate varies through the year, being lowest in October and rising to a peak in January and February.

Re-mating Frequency

Some does are re-mated post partum, that is to say immediately after they have given birth to a litter. If the plane of nutrition is sufficiently high, then the doe will satisfactorily carry one litter while suckling the first. If, however, the plane of nutrition is not sufficiently high, then there will be a considerable tendency for the unborn litter to be resorbed. A more usual practice is to re-mate the doe ten, fourteen or twenty-one days after the birth of the litter.

A twenty-one day re-mate cycle means that the average span between litters is fifty-two days, whereas the ten day re-mate gives a span between litters of forty-one days. This means, in effect, that on the twenty-one day re-mate system, almost exactly seven litters a year are possible, whilst on the shorter period nearly two more are obtained. However, the number of litters a year per doe is not generally the limiting factor on profitability. This is, rather, the number of young weaned per cage space. Doe replacements are not expensive to breed and slightly later weaning will give youngsters a better start.

Artificial Insemination

Artificial insemination in the rabbit has been developed to an advanced degree. In some of the extremely large farms in Europe it is the sole method used for getting does into kindle, and the results obtained with improving techniques are certainly the equal of natural matings. When a satisfactory system is developed and the numbers being artificially inseminated are large enough, then there can be time savings over natural mating systems – the ratio of bucks to does can also be reduced. The average quality of bucks used is also higher because fewer are needed. It must be said, however, that it is a skilled operation and usually requires two people working together.

The system involves the collection of semen from donor bucks in an artificial vagina. The semen is diluted and then injected into does which have been ovulated, in other words, injected with a drug which causes the eggs to be shed from the ovaries. Artificial insemination is not practised widely in this country and if a rabbit farmer is interested in using the system, he should quite definitely obtain professional guidance and instruction.

First Breeding

The age at which does are ready for first breeding is variable. The correct age will depend more upon the bodily development of the doe than on her age. In general, the smaller the mature size of the breed, the earlier the doe can be bred. Within the same size, better production is obtained from a doe if she is bred earlier rather than later.

It is sometimes advocated that the correct time to first breed a doe, provided she has been well reared, is when she weighs as little as sixty per cent of her adult weight. This is, however, too early and a level of seventy per cent should be taken as a more accurate indication of the ideal time. There is a disadvantage in leaving the doe later than this – for example, in the usual strains of New Zealand White, seventy per cent

Management and Stockmanship

weight will correspond to something like four months of age in well-grown animals.

The buck is usually bred later than the doe – eighty per cent adult weight is a good guide and he will probably be a month older than the average doe.

Palpation

It is desirable to test the doe for pregnancy as soon as possible after mating has taken place. There are pregnancy-diagnosing instruments available, but these are expensive and, if properly undertaken, palpation of the doe at about fourteen days is a perfectly reliable technique. Palpation, which is best learnt from an experienced person, involves feeling the embryos in both uteri of the doe. The sacrifice of a pregnant doe on the fourteenth day of her pregnancy, and a careful internal examination of the abdominal cavity, will give a clear idea of what to feel for.

The embryos are felt gently but firmly between fingers and thumb, the doe being held with the left hand (if the operator is right-handed) over the shoulders, or restrained by the ears with the head 'cupped'. The

Palpating.

Management and Stockmanship

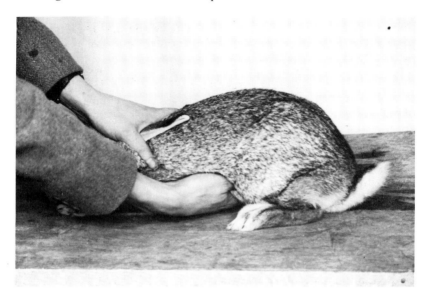

Palpating.

fingers must be pushed fairly deeply into the abdomen and the smaller faecal pellets should not be confused with the embryos. The embryos are to be found either forward, towards the rib cage, or much further back towards the rear of the animal. It is essential that the doe should be relaxed. A frightened doe with tense muscles makes the job more difficult and sometimes impossible.

KINDLING

Nest Boxes

The nest box and its management are very important for success. Unless it is a permanent feature of the doe's cage (the best solution) it must be put into position, in a clean and disinfected state and properly prepared, a few days before the doe is due to kindle.

If the box is a metal one, the use of an insert of soft wood, at least half an inch (one centimetre) thick, makes an excellent insulation, provided it fits reasonably well. Good insulation is essential and the nest box must be lined with material from which the doe can, with the addition of her own fur, make her nest. Coarse softwood sawdust or wood shavings as a

Management and Stockmanship

bottom layer are satisfactory, with a layer of wood shavings, shredded paper, dried hemp flakes or short straw on the top; provided there is sufficient this makes an excellent lining.

The doe may remove or eat some of the bedding and it is important, therefore, to ensure that there is ample at all times. Nest-building ability tends to be inherited and, because of its extreme importance, the quality of the doe's nest should be one of the criteria by which animals are judged for selection to produce future breeding stock. To help does, when the young have finished with the nest, any good loose fur can be removed, cleaned and baked to disinfect, and then used to 'top up' nests in the future. It is better, however, for good nest-building characteristics to be inbred.

The nest box should be inspected as soon as the doe has kindled – any dead young or mess should be removed and the litter counted and recorded. It is as well to distract the doe while the nest box is being examined. With an outside nest box the aperture can easily be blocked off from the inside.

Should any young be found outside the nest and chilled, they are best warmed thoroughly (95–98°F (35–37°C) is about the right temperature to keep them in) before being returned to the nest. It is remarkable how many apparently lifeless ones can recover.

Kindling Trouble

Does may neglect, scatter outside the nest box, or eat their newly-born young. There appear to be several reasons for this. The problem occurs more frequently in young does and may be due to stress induced by fear (of noise, vermin, and so on). Bad nutrition may be a cause and (much more frequently) lack of water is another. The doe will also eat new-born young which are injured at birth. There is no prevention for this, other than ensuring the doe is comfortable, has good food and, most important, water, and is kept in tranquil surroundings. A doe neglecting or damaging two litters should be culled. There is an inherited predisposition to good maternal characteristics which should always be recorded and used in selection. However, even in the best-managed rabbit units, using a fully intensive system, a nest box mortality of below fifteen per cent (including still births) is most improbable. It is the size of the remaining litter that is important.

Management and Stockmanship

DOES AND LITTERS

Young rabbits are born naked and blind in a warm nest prepared by the doe with bedding and her own fur. She will allow the young to suckle only once per day and even then for only a few minutes. Young rabbits do not have a fixed teat which is why a doe can rear more young than she has teats. The number of teats varies from six to twelve, with eight probably the most common.

The young suffer badly from cold and therefore a good nest contained in a proper nest box goes a long way towards a healthy weaned litter. The ideal temperature in the centre of the nest box is 86°F (30°C).

Young rabbits may start to leave the nest as early as the twelfth day after birth, although usually this happens slightly later. Much depends on the amount of milk the doe gives. Rabbits' milk is one of the richest of any animal, containing thirteen to fifteen per cent protein, and ten to twelve per cent fat. The early growth of the young rabbits is, therefore, very considerable. By the twentieth day or so, the young rabbits will be eating some of the doe's food and the earlier this happens the more indicative it is of poor milk yield. The young should have the best possible food at this age, although almost invariably a special diet is not introduced.

Litter size is variable, both amongst different breeds and within a breed or strain. The smaller the breed the smaller the litter size tends to be. There are variations also in the sizes of the litters (that is, between the first, second, third etc.) The first litter of a doe is on average smaller than the second litter, and the weight of the youngsters also tends to be lower. The second litter will also be smaller than subsequent ones.

At this stage of the production cycle there is nothing that the farmer can do, except to inspect the nest boxes and ensure a supply of food and water and relative freedom from undue stress.

Milk Production

The amount of milk produced by the doe is the major factor in the growth of the youngsters in the nest box. It is important to do whatever possible to stimulate the supply of the milk, either by the initial selection of the future breeding stock, or by management.

There is no doubt at all that does first mated at too early an age (for example, at 100 days), give less milk in their first litters, and in subsequent litters, than those mated on average three weeks later. The best possible food, rich in protein and fat, is important for milk production, as is the water supply – a doe with a growing litter can drink a considerable amount. The amount of milk produced is also severely

Management and Stockmanship

affected by excess heat and the temperature in the rabbit house should not exceed 59°F (15°C). Milk production is greatly reduced if the temperature rises to 86°F (30°C).

Milk yield is, to a fairly large extent, influenced by quantitative genetic characteristics and it can, therefore, be improved by selection. Good milking quality in the doe can be assessed by the growth of her young to twenty-one days. Litter weight at twenty-one days is a good criterion, therefore, for selection of mothers for future breeding stock.

The full genetically-determined milk yield can only be produced if the nutrition of the doe is of a sufficiently high standard. Usually the milk supply for the second litter is substantially better than that for the first. If this is not the case, then it should be suspected that the nutrient supply is insufficient and that the doe has drawn on her bodily reserves for the first litter and has nothing left for the second.

The age of the doe, or rather the size of her litters, will have quite a considerable influence. Provided that the nutrient supply is satisfactory, the milk yield will increase in rapidly-producing does up to about the sixth or seventh litter. It will then stay at approximately the same level until the tenth or twelfth litter when, if the doe continues to breed, it will start to decrease.

The amount of milk yielded is stimulated by the suckling of the young, but only up to a certain level. It is probable that this level is reached with six or seven young suckling, and after this there is no further pro rata gain.

Finally, one of the early indications of a lacking milk yield in the doe is the premature leaving of the nest by the young animals and their early consumption of the doe's food. A tendency to eat solid food before the end of the second week of its life, indicates that the young rabbit is over-hungry.

Fostering

Fostering involves the transfer of young between does and is done when one doe has an excess of young and another has too few. It is usually best effected in the first few days of life (never more than about five) and it is most successful when the two sets of young are born on the same day. Only young from a completely healthy doe should be fostered.

One of the problems associated with this practice is that the parentage tends to be lost. For this reason, unless the young are suitably marked by a small patch of dye, fostering should not be used when it is important to know the true parentage of all young.

Fostering also, to some extent, modifies the records of food conver-

Management and Stockmanship

sion and so on. When records are kept, then all fostering should be included on them. A plus or minus sign (with the number of young fostered into, or taken from, a litter) added to the recorded litter size is usually adequate.

Rejection by a doe of fostered young is unusual provided always that the doe is not disturbed too much when the young are introduced. As the doe suckles her young for only a few minutes a day, the closing-off of the nest box for a period, in order to carry out the transfer, is of no consequence at all.

WEANING

The best method of accomplishing weaning has been a subject of much argument. It is believed by some that the stress induced by weaning and moving youngsters to a fresh environment has an adverse effect on their growth. Some other experimental results indicate that under certain circumstances, this may not be so.

Two solutions to possible weaning stress have been tried. The first is to install a double cage system in which a small aperture allows the youngsters to move into a second cage (which becomes their fattening cage) whilst they are still with the doe. At weaning time the aperture is closed and the youngsters are left in the second cage. This system requires the initial manufacture of the special cages and increases cage requirements with early re-mate systems.

The second solution lies in the use of anti-stress injections of antibiotics – any success these may have will possibly be due to them having prevented disease rather than stress.

In a well-managed rabbit unit such refinements are rarely necessary and a straightforward moving of the youngsters to their fattening quarters is the best solution.

The age at which weaning can take place is dependent on several points. Firstly the litter cannot be left with a doe which is about to kindle. Hence with a post partum re-mate system the young must of necessity be weaned at twenty-seven or twenty-eight days. At this time, the doe's milk (especially in the case of a good milking doe) supplies a quarter to a fifth of the total food of the young. The very rich milk supplement is certainly a help. None the less, a number of successful breeders do wean at this time. There is usually a minimum weight of about 1lb (0.5kg) for a young New Zealand White below which it should not be weaned.

In a ten day re-mate system, the young are weaned at five weeks of age. There is no point in weaning later than six in any system, for the

Management and Stockmanship

milk supply of the doe will have ceased by then.

On weaning, the young are best placed in family groups, as far as possible, ideally six or eight to a pen. They can remain in such a pen until slaughter at up to twelve weeks, or be removed to other pens at that age (in small groups in the case of does, or singly in the case of bucks) to be retained as breeders.

GROWING STOCK

Does selected as future breeding stock can be kept together until they near breeding age. Bucks to be retained after about twelve weeks of age must, however, be housed separately or fighting will occur.

With the does, there is the possibility of pseudo-pregnancy and this can only really be avoided by separating the animals to individual growing cages. Pseudo-pregnancy is as its name implies and is produced where there is no fertilisation of the eggs that are released as a direct result of sexual stimulation of the doe. The condition occurs if a sterile buck mates a doe, or if, for some other reason, the eggs are not fertilised. It also often occurs when sexually mature does mount each other.

Pseudo-pregnancy lasts for fifteen to eighteen days and its termination is indicated by the doe appearing to make a nest. If she does this then it is safe to assume that she was pseudo-pregnant and she should be mated on the eighteenth day. At this point she is more receptive and fertile than usual.

There is one major problem which arises when growing on young animals for future breeding – the object is to get the animals well-grown but not over-fat. Any amount of fat is deleterious to breeding and, as a consequence, it is necessary to restrict in some way the rations of the animals. This can lead to further difficulties – if it is done by a simple limitation of the high-energy food used for breeding stock and weaners, the appetite will not be satisfied and fur eating may occur. Fur eating produces hair balls in the stomach and can damage the pelt of meat animals, resulting in a penalty from the processor. The only way to combat fur eating is to supplement the rations of the animal with a bulky fibrous food such as hay.

GROWTH AND DEVELOPMENT

The rate of growth (increase in weight) and development (the change in the proportions of the body, both organs and tissue) are controlled by a number of factors.

Management and Stockmanship

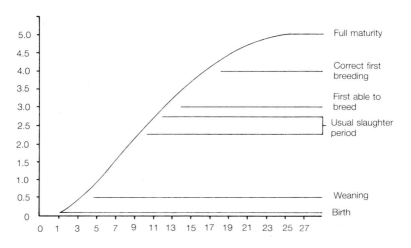

Simplified growth curve.

The typical growth curve, plotting weight against age, is shown above. The figures are fairly typical of the average New Zealand White.

The initial birth weight tends to be dependent on the doe's size and, to a slight extent, on the number of young in the litter. In the early stages, growth is very rapid and it is at this time that the best food conversion rates are achieved. At about the beginning of puberty, the rate of growth declines and after six months or so no further growth is made. The basic growth pattern is controlled by inheritance, but the animal's food and its environment have a vast influence and effect on this basic pattern.

The aim in commercial rabbit farming is to produce a good carcass at an early age using the lowest amount of food. The most economical way to do this is to encourage the fastest growth possible.

CULLING

Culling is the routine and systematic elimination of any animals which are unlikely to be of value in the unit, and it must be done as soon as this

Management and Stockmanship

lack of present or future value is determined. The rate at which general improvement in the herd takes place depends to a great extent on the severity of the culling. There is no doubt at all that there is far less culling in the UK than in France, Italy and Spain, countries where rabbit husbandry is often more fully intensive. Systematic and skilled operation of this practice will probably have the greatest effect in the shortest time on the profitability of a rabbit herd.

There are several reasons why an animal should be culled. The first is ill health. Not only can an unhealthy animal never be profitable (because of the loss of its own production) but it may also be the cause of further loss as it infects others.

A number of undesirable conditions are directly inherited, or there can be an inherited predisposition towards them. Most of these conditions are recessive in nature and, therefore, are not seen to the fullest extent (or at all) in the parent stock. Malocclusion of the incisor teeth and yellow fat are examples of this. If an animal is found to be affected, it follows that both its parents must possess the genes for the characteristic and, therefore, they should no longer be used for the production of future breeding stock. 'Sore hocks' also appears to run in families and animals suffering from severe cases of this condition should be culled. Unless the desirable characteristics of certain rabbits are so superlative that the animals cannot be overlooked, it is important not to select stock from any of the progeny of the parents of affected animals.

It is very difficult to give any precise indications as to the amount of culling that should take place. For one thing, it will depend to an extent on the quality of the herd. If the average quality of the herd is poor, and variability of quality between the animals in the unit great, more culling needs to be done. In the same way, the larger the number of animals the more extensive the culling must be. In the best rabbit farms, it is likely that the doe replacement will probably be about 100 per cent a year giving the average life of all breeding does as about eighteen months.

The performance of bucks is checked on a continuing basis. After ten or twelve weeks of use a buck performance analysis (*see* page 116) will indicate how he is shaping up in overall performance – by that time he should have served fifty or sixty does. Provided the farmer has adopted the policy of having a greater number of bucks than is strictly necessary, he is in a position to cull the worst.

A word of caution should be included here – it is important to remember that like should always be compared to like. Conception rates, for example, will vary through the year and it would be wrong to compare the performance of one buck in the period September to November, with the performance of another in January to March.

Management and Stockmanship

Whilst culling should be a continuous process, it should be more severe at some times of the year than at others. The successful farm will attempt to have a regular production of young animals for sale throughout the year. Poor farms have a far higher production of young during the summer months (when demand is lower), than in the winter months, when the demand always tends to exceed the supply. Therefore, culling to reduce the amount of breeding stock for production in the summer, and the replacement of the does for increased production in winter, represents a desirable pattern to follow. A thorough check of all animals being bred in early spring is not a bad practice.

Whilst the records themselves should be the basis of culling, handling the animal can also give an indication of health, condition, bodily conformation and so on.

IDENTIFICATION

In all cases in which long-term records are used for the selection and improvement of the stock, the animals themselves must be individually identified. The only exception to this occurs when the food consumption of the litters of selected does is to be recorded, and the youngsters themselves are not to be retained after slaughter age. In such cases, the litters can be identified by the cage numbers only.

There are three methods of marking rabbits for permanent identification:

1. Identification rings of various sizes for the different breeds of rabbit – this method is never normally used by the commercial farmer. Official rings, used for exhibition purposes are obtainable from the British Rabbit Council.
2. The tattooing of a number in the ear. A tattooing punch with changeable numbers is used to produce very small holes which are filled with special marking ink. Numbers and or letters are used, usually in the left ear for bucks and in the other ear for does. Before punching, the ear is cleaned with industrial alcohol, the inside of the ear is then punched and the ink rubbed in immediately.

When tattooing is carried out single-handed, a tattooing box with adjustable floor and one adjustable end can be used to make the work easier. The overall size of the box is usually 20in × 8in × 8in (50cm × 20cm × 20cm) and it should have a hinged lid with a hole towards one end through which the ears can protrude. The floor is raised or lowered for different sizes of animal and the end moved back and forth. In the case of very young rabbits the ears of which are too small for a tattoo

Management and Stockmanship

punch, a tattoo needle can be used. Sometimes a very simple temporary system of a few holes in one or other ear is used for early identification, followed by the proper punching at a later stage.

The process of tattooing is rather laborious and is, therefore, only used when it is essential for the method of identification to be absolutely certain.

3. The ear tag or ear stud – the most commonly-used method. There are several varieties of these and they can be obtained serially numbered, usually up to 10,000. Some ear tags require closing pliers, whilst others can be fixed manually. Unfortunately, the ear tags can be torn out, causing some damage to the ear and also loss of identification. When ear tags are fixed in the ear (usually about half way between tip and base) some space should be allowed for any increase in the size of the ear.

It must always be remembered that whilst proper identification is absolutely essential when accurate recording is to be carried out, animals are sometimes needlessly marked by farmers.

WEIGHING

Weight is an important indication of health in the rabbit when correlated in a known strain of animals. It is also a better guide to the correct time for first breeding than age alone. The weighing of litters is essential, too, if food conversion rates are to be established for the selection of future breeding stock. A good weighing machine is, therefore, of great importance.

The ideal weighing machine is a dial machine with a tare adjustment and a dash pot damper. The tare adjustment means that a satisfactory container can be placed on the platform and the dial adjusted to zero to allow direct reading. The dash pot damper makes the pointer come to rest quickly despite movement of any of the animals. The container needs to be kept clean and disinfected regularly. A 22lb (10kg) machine is usually sufficient but a slightly larger one is probably better.

A piece of equipment like this is rather expensive and many rabbit farmers use a spring balance with a container hung from the hook. Rather than simply holding this, it is better to arrange that the body of the balance is mounted or supported rigidly.

Some standard weights should be kept for a regular check of the readings of the weighing machine.

Management and Stockmanship

MANAGEMENT PLANNING

One of the secrets of success of all animal enterprises is the development by the operator of a routine, to be strictly observed. This improves the efficiency with which operations are carried out and it also ensures that nothing is missed. It is important, particularly when the enterprise is first started, to plan the routine in some detail – writing it down helps not only to clarify it in the mind of the operator, but also to remind him of it afterwards. One of the effects of sticking to the same daily procedure is that the animals themselves are less likely to suffer stress than if there are variations. Once the routine has been written down, a weekly planning sheet can be devised. The tasks that have to be undertaken are:

- Mating does.
- Palpating does for pregnancy.
- Re-mating does not in kindle.
- Placing, preparing and inspecting nest boxes.
- Weaning and weighing.
- Marking for identification.
- Selecting future breeding stock.
- All routine tasks such as feeding and cleaning.

These tasks all relate to different animals, and once a certain number (fairly small) of breeding does has been reached, a great deal of time will be wasted if the farmer does not develop some system of identifying easily and accurately what tasks have to be done for which animals.

Tasks should be done in groups – all animals for weaning can be weaned on the same day each week, all matings done on particular days and so on. The use of the matings register, if kept in full, goes some way to acting as a permanent reminder; but the planning itself must be done initially.

The amount of time taken to complete all the work necessary in a rabbit unit is variable. It depends not only on the skill and experience of the farmer, but also on a number of other factors – the size of the unit, its layout, the design of the equipment and the type of labour-saving aids employed, for example.

An analysis of some of the best fully intensive farms shows that the labour necessary for each breeding doe (for the production of meat animals) can be reduced to some seven hours per year. In the same way, the following group estimates are achievable in good conditions with good routines:

Management and Stockmanship

	Hours per 100 breeding does per year
Matings and re-matings	120
Nest preparation and inspection	110
Doe palpating	50
Weaning, weighing and marking	125
Feeding and general	130
Cleaning and disinfecting	90
Detailed recording and analysis	110

These are good figures and, as long as the tasks are all being properly carried out, would indicate that the unit is being run efficiently.

Targets of Management

Profit is the sole aim of a rabbit farming enterprise, and whether or not the unit makes that profit is the only way of measuring its success. However, profit cannot easily be measured on an ongoing basis and some form of target or standard, against which the farmer can measure the efficiency of his management and the degree of its success, is, therefore, appropriate. Clearly, it is necessary to complete and analyse the records of the unit in order to compare one's management with these standards. This is discussed more fully later.

The matching of production statistics in a unit against such targets also tends to pinpoint that unit's weaknesses – in this way the reasons for the weaknesses are more easily established and eliminated.

Some good production figures for the different animal characteristics that might be measured (some of these may be contradictory) are as follows:

Conception rate   80%
Average period between litters   52 days
Average weight at weaning   1.3lb (0.6kg)
Number young weaned per doe/year   50
Food conversion rate   3
Pre-weaning mortality   13%
Weaning to slaughter mortality   2%
Feed as a percentage of all costs
(a lower percentage means other costs are too high)   75%

Management and Stockmanship

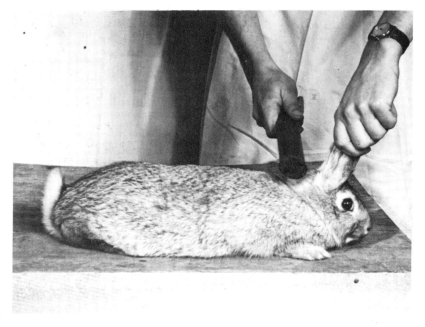

Killing by blow.

KILLING

The great majority of farmed rabbits are sold live to a processor and passed through a slaughter process which includes electric stunning and decapitation. There are, however, occasions on which the rabbit farmer himself will need to kill animals quickly and painlessly.

There are two methods used. The first is the blow method in which the animal is killed with a heavy stick 12–18in (30–40cm) long, being hit at the base of the skull. The blow should not strike the shoulders for it might then not be effective, and the carcass could be damaged. This method should be used only by those unable to learn and use the second method – dislocation of the neck.

The animal is held by the hind legs in the left hand, whilst the fingers of the right hand pass under the jaw – the thumb is placed over the back of the head. The hind legs are held securely and the head is lifted high at the neck so that the jaw line is at right angles to the spine. If too much force is used by an inexperienced person on a young rabbit the head may

Management and Stockmanship

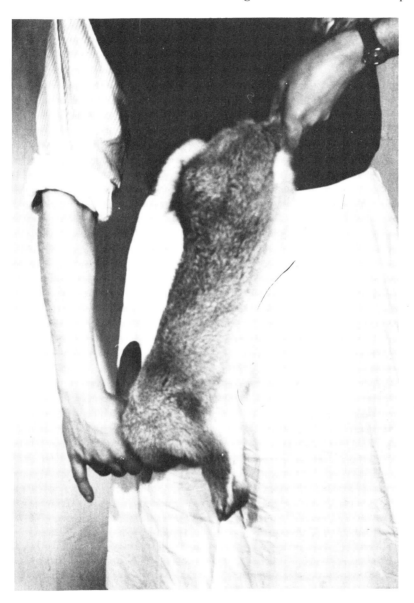

Killing by dislocation.

Management and Stockmanship

be pulled off. A sudden 'give' will be felt when the neck is dislocated. A demonstration and lesson from an experienced farmer is the ideal way of learning the knack.

If a live rabbit is held with the head hanging downwards the ears will be held tightly back – as soon as it is dead, the ears fall forwards.

OVERGROWN TOE NAILS

The toe nails of rabbits kept on mesh floors become overgrown as they are not worn down. It is, therefore, necessary to clip them from time to time. Some farmers do this as a regular routine, for example, when youngsters are removed from the doe or when palpating. The toe nail can be clipped to within about 1/4in (6mm) of the quick, which can be seen as a shadow inside the nail. Cutting into this may cause pain and bleeding, so great care should be taken. A sharp pair of heavy-duty nail cutters (or nail clippers) is ideal for this job, the curved blade scissor-type being the best.

MANURE DISPOSAL

The breeding doe and her followers (all her young in the year to slaughter, together with a replacement or two) are likely to produce about 550lb (250kg) of manure during the year. Manure averages around 8cwt per cubic yard in volume (500–600kg per cubic metre) and this means that each breeding doe with followers is going to produce about two-thirds of a cubic yard (half a cubic metre) of manure per year.

Provided that ventilation is adequate, that some steps are taken to ensure that urine can drain away and that the houses are proofed against rats and mice, the manure can be left under the cages for a month or two, or even longer. Where the manure is easily removed, however, many farmers prefer to clean it out at more frequent intervals. Sprinkling super-phosphate on the manure helps to prevent the nitrogen in the urine from being leached away and improves the quality of the manure as a fertiliser. It also assists in the prevention of excess ammonia smell. Quicklime has the same effect, but is more unpleasant to use and can, in certain cases, be harmful to the animals.

The manure can be considered as a valuable additional source of revenue, but a good deal of care has to be given to its handling if this is to be realised.

There is considerable interest, both in America and in many parts of Europe, in the use of different types of worms (particularly *Eisenia foetida*), for the conversion of manure into an excellent compost. The advantages are that it greatly assists in keeping down smells and the maggots of flies, as well as successfully converting the manure. The worms are, however, killed by vermin, heat, cold, mould, excess urine and so on. A fair amount of expertise is required, therefore, for the use of worms and it may be better, if the manure is to be sold, to dry, crush and pack it without attempting to convert it to compost. A local garden centre may offer a suitable market.

5 Other Aspects of the Rabbit

Commercial rabbit farming is concerned with the production of meat animals. However, there are other aspects that are of interest to the farmer, although they may, perhaps, not be suitable for business purposes.

ANGORA RABBITS

Angoras have been farmed for their wool for a very long time. In the 1970s and 1980s there has been an increasing world-wide demand for the wool and the prices have gone up considerably.
It must be said immediately that angora wool is a luxury article which is subject to the vagaries of fashion. The price is, therefore, likely to fluctuate wildly. Furthermore, there is such interest in angora wool farming throughout the world at the present time, that it is very likely that there will be over-production – it is probable that a substantial fall in the price will follow in the next few years.
Angora wool farmers have always selected their breeding stock on the basis of each rabbit's performance as a wooler – indeed, this has been done to such effect that the wool yield of does has increased probably five-fold in the last 100 years. It used to be as little as 0.5lb (250g) per year, but can now average over 2.75lb (1250g) per year for the best. The heritability of wool yield is probably the highest of any characteristic in the rabbit.
In the commercial enterprises in both France and Germany the males are usually not kept for wool production as their yield is a good deal less than the females. The reproductive abilities of the Angora are as little as fifty per cent as efficient as those of short-haired breeds. It is for this reason that only about one doe in twenty is used for producing replacement animals. The average life of a wooling Angora will vary in length between three and four years.
There are a number of different types of Angora. The English exhibition Angora yields only half the wool given by either the French or the German types and the continental types are, therefore, the only ones used in commercial Angora farming.
The production of the best yields of wool depends to a great extent on

Other Aspects of the Rabbit

A part-sheared Angora.

the quality of the feeding. Production on non-concentrated feeding stuffs can be as low as half that achieved when the animals are correctly fed on pelleted feeding stuffs. Food costs vary enormously but tend to be at best thirty per cent of revenue, and at worst sixty per cent.

Wool is usually harvested four times a year and is either sheared (the normal practice in Germany) or plucked (more common in France). The latter method produces the best result, but it is time-consuming.

The time requirement for wooling Angoras varies, in the most efficient enterprises, between four and five hours per animal per year.

At the present time, there is virtually no production industry in angora wool in the UK, although there is a great deal of interest. Angora husbandry has possibilities, but it is certainly one of the more speculative aspects of rabbit farming.

FUR RABBITS

In the 1920s a rabbit fur industry developed in the UK, and in other parts of Europe. In the latter part of that decade pelts were selling for a price which at today's values would be very high indeed. It is true that in

Other Aspects of the Rabbit

order to produce the best pelts the animals had to be kept for six months on average. The sale of the pelt was considered, however, to cover the cost of food and overheads, and the value of the carcass made the profit. Feeding was traditional with roughages and hay, and perhaps the occasional cereal. However, with the depression of the early 1930s, and the development of Russian exports of furs to obtain foreign currency, the industry was quite killed off.

From time to time attempts are made to re-establish rabbit fur production, but with little success. At present-day prices the age to which the animals have to be kept in order to produce a good quality fur, means that the industry is not economically viable. The value of the pelt of a young rabbit of ten or twelve weeks of age is very low. Indeed, at the present time they are often quite unsaleable. The fur itself, at the age at which the animals are killed, is not free of moult and is not well-grown enough. It is also of a quality too poor for felt making and if it is sold for a few pence (in large quantities) it usually ends up for cutting, glue-making or in fertiliser.

There are a few smaller-scale rabbit farmers who have developed some fur activities from their own stock. In these cases, however, the fur is used for novelty items of relatively low value. The skins have to be taken more carefully than in normal slaughter practices and must be dried. Although they can be home-cured, it is usually best to have them professionally treated to produce a good result. They can then be cut and sewn into a wide variety of goods – there is also a small sale of such items through handicraft shops.

SALE OF LIVE ANIMALS

Live animals can be sold for purposes other than slaughter for meat. Firstly, there is a minor rabbit enterprise in the sale of animals for domestic use. Much of this is carried on through pet shops and is an activity that should be of no interest to the rabbit farmer.

Secondly, there is the question of the sale of breeding stock for commercial meat production. There are two points of view concerning this. One says that the cost of the time involved in showing potential purchasers the animals outweighs the enhanced price obtained for breeding stock. There is also the point that producing future breeding stock is more expensive to do because it has to be accomplished with the necessary recording and with improved lines. The opposite view is that if a specialist breeder sets up to breed first-class breeding stock for sale, then it can be a viable and reasonable operation.

It cannot be too frequently stressed that the rabbit industry and

Other Aspects of the Rabbit

people coming into it have often been cheated by people whose interest lies solely in the making of money, no matter how dishonestly. If, therefore, a rabbit producer sets out to establish himself as a seller of breeding stock, then he should adopt the following strict policy rules:

1. He should keep a meat production unit of a reasonable size at the same time.
2. He should not buy in stock for re-sale.
3. He should take steps to become an accredited breeder of the Commercial Rabbit Association.

The sale of breeding stock is not a matter for the relative newcomer to the rabbit industry.

Laboratory Rabbits

In the past there was an excellent outlet for the commercial rabbit farmer in the market of laboratory animals. The prices obtained for these tended to be far higher than those obtained for meat animals, and they often exceeded even the prices paid for breeding stock. With increasing pressure for a reduction in the use of rabbits (and other animals) in laboratories and the development of other means of testing, the market for laboratory animals is diminishing.

The price for these animals is high and they have to be of a quality that warrants this. Furthermore, the recent introduction of the Animals Scientific Procedures Act (1986) means that all producers of laboratory animals will have to register under the Home Office and follow strict guidelines laid down by them. It is clear, therefore, that this side of rabbit production is very specialised and best left to the extremely experienced farmer.

THE RABBIT FANCY

In the UK (and in all European countries and America), there is a well-developed Rabbit Fancy. Many thousands of fanciers produce a number of different breeds to the numerous standards that have been drawn up by National Specialist Clubs (of which there are over forty in the UK), and published by the British Rabbit Council.

The animals are exhibited at shows that are held throughout the country. These shows (1,500 or so each year), vary in size from under a hundred rabbits, to well over a thousand – the number of exhibitors can be just a few, or up to five or six hundred.

Other Aspects of the Rabbit

The prize money in classes in which the animals are exhibited is small and only very few of the most successful exhibitors might possibly break even – when the costs are all included it is a hobby (indeed for many a total dedication!) which is quite expensive to pursue.

The majority of the shows are one-day events, but there are some larger shows that go on for two days. There may be as many as 400 different classes which are either straight breed classes, or duplicate classes. The straight breed class is confined to a particular breed or a particular colour of breed and limited, possibly by age and/or sex. The duplicate class is for animals drawn from different straight breed classes.

All shows are held under British Rabbit Council rules. The Council is an organisation to which almost all of the clubs in the country (well over 400) are affiliated. The whole subject of the exhibition of rabbits is extremely interesting and a visit to a good event (many Agricultural Societies have rabbit sections at their shows) is most instructive and enjoyable.

Some mention should also be made of the carcass classes which are occasionally held. The basis of such classes is that the animals are slaughtered at a central station and detailed weighings are undertaken. The dress out percentages are then calculated and the carcass judged as to its conformation and quality. Such carcass classes do not perhaps help to improve the quality of breeding stock, but they are most educational for producers, leading to a more thorough appreciation of conformation in the live rabbit and its relation to quality in the carcass.

6 Disease

It is impossible within the confines of a book such as this to do more than indicate the signs of good health and of disease, and to attempt to outline some of the points which the rabbit farmer must consider if he is to be successful in preventing large disease losses.

Disease within the unit is one of the most important matters to which the rabbit farmer must turn his attention; but it is also a very specialised subject and it is therefore essential that, when trouble does arise, he should obtain professional assistance as soon as possible. It cannot be stressed too often that attending immediately to the first signs of trouble is vital. A disease problem tackled straight away is a far easier problem to deal with than one that is left in the hope that it will go away.

THE CHANGING PROBLEM

At one time the problem in rabbit units (which tended to be small and nowhere near as intensive as they are today), consisted of a number of fairly definite diseases which were, in most cases, susceptible to diagnosis and often to treatment.

In today's intensive units this no longer applies. Disease tends to be chronic, with the animals performing at a lower level than they should. Some people blame the stress of intensive housing and intensive production – certainly, if animals are to be farmed profitably, and if their levels of production are to be increased to the best possible limits, then this 'stress-induced' ill health has to be prevented. The disease problem has changed a good deal over the years, and now it is much more closely related to the buildings (and their ventilation and so on), to feeding, to management practices and to the general welfare of the animals, than it was in the past.

Diseases are the result of excessively large populations of bacteria, viruses, protozoan parasites, and other disease organisms building up to such an extent that the animal's own protective system cannot deal with them. It must never be forgotten that there are two aspects involved – the build-up of disease organisms and the inability of the animal's defences to protect it. Poor conditions of environment and bad

Disease

nutrition, added to what might be termed the biological exhaustion of the animal, are likely to lead to the animal becoming diseased. It is for this reason that not only must the rabbit farmer have the most efficient hygiene and disease prevention practices possible, but he must also pay the greatest attention to producing the best possible condition in the animals.

There are a number of causes of disease, and different conditions attack animals at varying ages. Furthermore, each unit will have a dissimilar incidence of the various diseases in each age group. The table below gives a generalised idea of the main groups of disease and at what ages they are most likely to kill the animal that is affected.

There are two important areas in which the rabbit farmer must make himself proficient. Firstly, he must devise a system of examination of his rabbits. The random examination of an animal will lead to missed observations, whereas, if the examination is systematic, and this systematic approach always practised, it is much more likely that all significant points will be recognised.

Secondly, the rabbit farmer must make himself thoroughly conversant with the signs of good health and the signs of disease. A healthy animal will be alert with vigorous movements, actively feeding and drinking, have clear bright eyes, typical healthy posture,

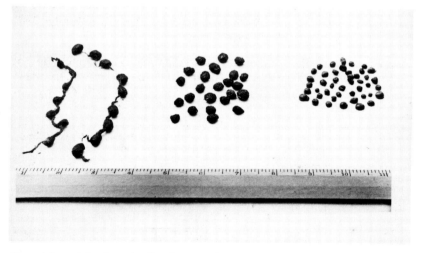

From left to right: Faecal pellets from a sick animal, pellets from a normal rabbit and cophragous pellets.

Disease

clean fur and skin, and will often be seen grooming and so on. The faecal pellets are firm, of correct size, not strung together and of normal colour.

Perhaps even more important are the signs of ill health. There are four main areas to observe which can indicate when all is not well:

Appearance
Discharges
Activities
Others

Signs to look out for are as follows:

Appearance: staring dull coat; unusual, strange or huddled posture; swollen stomach; wounds and sores; lumps, bare patches on body, legs or feet; stained or matted fur; protruding teeth; swellings.
Discharges: which may be blood, mucus (thin and watery or thick and purulent) and so on, from the eyes, ears, nose, mouth, anus, sexual orifices or, indeed, from the body itself.
Activities: general lethargy; scratching at the ears or the body; shaking the head unusually; wiping the nose with the fore paws; sneezing or grinding of the teeth; breathing rapidly, heavily or noisily; moving unusually – limping or dragging its feet; paralysis.
Others: loss of weight or condition; constipation; not feeding or drinking; diarrhoea – the nature of the faeces is always an important indication of the health of the animal.

A good stockman will train himself to recognise immediately any of these indications. As soon as any are detected then he must take action to remove the animal or animals concerned to an isolated area for

	1–7 days	1–4 weeks	5–12 weeks	Growing stock	Adult	Total
Gastro-intestinal diseases	–	6	16	6	6	34
Respiratory diseases	1	9	8	5	7	30
Reproductive disease (and deaths due to poor mothering)	28	1	–	–	1	30
Others	1	2	1	1	1	6
Total	30	18	25	12	15	100

Percentages of all deaths at different ages.

Disease

quarantine or humanely destroy them, if he is certain about the cause of the trouble. If he is not certain then he must take steps to find out the cause, how to prevent any spread of the condition and its treatment. It is not usually economically justifiable to treat seriously ill animals. After the removal of the animal, the cage and equipment with which it was in contact must be thoroughly cleaned and disinfected.

The more intensive any form of livestock husbandry becomes, with a greater concentration of animals, and, indeed, the larger the unit, the more likely the growth of disease will be.

The disease problem must be tackled by preventive measures and the four most important elements in preventing disease are:

1. The selection and maintenance of first class stock.
2. The best nutrition, feeding and watering possible.
3. The prevention of the initial introduction of disease or sources of disease, and discouraging its spread and development.
4. Hygiene.

INTRODUCTION AND SPREAD OF DISEASE

Disease is introduced into a rabbit unit in a number of different ways. The most usual is probably by the introduction of new animals, or by the return of animals which have been in contact with diseased ones outside the unit. There are other ways too. For example, disease can be brought in by vermin which may move from house to house; myxomatosis can be spread by the mosquito (although it is more usually spread by the rabbit flea); and flies can carry disease over surprisingly large areas. The workers in the unit can also bring disease in with them.

Once introduced, disease is spread within a rabbit unit in an even greater variety of ways. In any rabbit unit there is considerable air contamination by disease organisms – respiratory diseases are spread easily this way. Others are spread by contact, for example, rabbit spirochaetosis, passed on during mating. The doe suffering infection with coccidia will pass on the disease to her young – coccidia are invariably present in the rabbit unit and a faecal pellet of an apparently healthy animal will usually contain sufficient coccidia to infect a young animal seriously. The movement of equipment such as feeders, nest boxes and so on also aids the spread of disease. Sometimes the worst distributor of disease is the human worker in the rabbit unit. By lack of attention to hygiene he can, for example, spread infection through all the does he palpates at one time. This is why it is so important for the operator always to examine thoroughly any animal he handles, and each

Disease

time he does so, to make sure there is no sign of disease which might then be passed on.

Hygiene involves not only preventing the introduction of disease organisms but also all those measures necessary to aid in discouraging any that do exist from spreading further.

FEAR AND OTHER STRESSES

The rabbit is a highly excitable animal and suffers considerable stress when subjected to any stimulus which causes fear – fear can be brought about by, for example, unaccustomed noise, the appearance of strangers in the form of animals or man (particularly the latter), a sudden and unexpected movement, and so on. Stress is also the result of any environmental effect which causes discomfort – fumes, smells, dust, undue heat or cold, draughts, exposure to the elements can all be to blame. There are also the physiological stresses, such as gestation and lactation.

The extent to which these stresses undermine the health of the animal has in the past been greatly underestimated. All the actions of the rabbit farmer must be directed towards reducing such stresses to the lowest possible level.

There is sometimes confusion in the matter of noise and its effect on the rabbit, and the way that humans can cause panic and thus harm. It is when the noise is, to the rabbit, inexplicable and when the appearance is not recognised that harm can result – fear is not caused simply by the presence of either noise or of a human being. It is for this reason that a good stockman will constantly speak quietly to his animals, and will avoid making sudden movements and appearances in front of them.

CLEANING AND DISINFECTION

Part of the hygiene programme in any rabbit unit involves the thorough cleaning and disinfection of cages and equipment – it is quite useless to attempt to disinfect any item unless it has first been thoroughly cleaned. One of the problems of the cleaning of mesh cages is that the cleaning can sometimes damage the galvanizing, which results in a surface that is unsatisfactory for stock.

Disease

Cleaning

It is more difficult to clean cages and equipment satisfactorily when they are actually installed, although this is almost invariably the way it is done. On the Continent, the larger commercial enterprises have a system of removing the cages (at certain times in the production cycle), to a special area for correct cleaning and disinfection. High-pressure hoses and/or pressure steam-cleaning are then used to do a very efficient job. In such cases it is essential to design the caging and equipment with this practice in mind.

More basic cleaning can be accomplished by scrubbing with a good detergent cleaner, although the removal of fluff and other such material should be completed before this is done.

Disinfection

There are no universal disinfectants, and some of those available are not suitable for use where animals are likely to come into contact with any residue. The only suitable disinfectants are those on the Ministry of Agriculture approved list. In all cases the manufacturer's instructions should be strictly followed.

Coccidia can only be dealt with satisfactorily by a ten per cent ammonia solution, and even this must be allowed some time to be effective.

In general, disinfectants used hot are more active than when used cold, but it is important to make sure that the liquid can be heated without causing damage.

THE RULES OF HYGIENE

Hygiene in the rabbit farm must involve the strictest cleanliness of the environment, of the housing in which the animals are kept and of the equipment they use. It also involves the strictest attention paid by the operator to his own hygiene, and the supply of clean and uncontaminated air, water and food.

Hygiene cannot be separated from any other element of management. The practices of hygiene must form an integral part of the work of the rabbit unit. It is impossible to formulate with any degree of precision set rules that are applicable to all circumstances. There are, however, some which are universal:

1. Prevent the entry to the rabbit unit (or to any associated storage or

Disease

work room) of any form of vermin, birds, domestic animals, insects and so on.

2. Restrict as far as possible human visitors and ensure that they observe all the rules of hygiene. Visitors who have recently been on other rabbit farms are a particular risk.

3. Ensure food is stored safely, in dry, vermin-proof conditions, and that it is not contaminated in any way.

4. Make certain that all animals arriving at the rabbit unit are placed in quarantine away from the main unit for a period of three or more weeks.

5. Ensure that all animals showing any sign of ill health are removed from the production unit without delay.

6. Ensure that all dead animals are safely disposed of.

7. Ensure that the interiors of buildings, cages, equipment, floors and so on are thoroughly cleaned, and that the word 'thoroughly' is firmly borne in mind by the cleaners.

8. Ensure that rabbits cannot come into contact with faeces, either their own or other animals', and that when the manure is removed it is done so safely.

9. Ensure that watering systems do not leak and that they function efficiently.

10. Ensure that the ventilation system is sufficient to remove fumes and dust, and to keep the environment in the best possible state.

DISEASE BUILD-UP

Regardless of the standards of hygiene and the preventive measures that are taken, sooner or later, with the present state of our knowledge of rabbit husbandry, the number of disease organisms present in the rabbit unit will reach an unacceptable level. The incidence of disease and subsequent loss of production will gradually increase as a result. The only way in which this problem can be tackled is to eliminate all the animals from the rabbit house (or at least from a section of it which can be physically isolated), carry out thorough cleaning and disinfection and give the house a period of rest to allow all disease organisms to become eliminated.

It is a serious mistake to re-establish the house with the same breeding stock that were taken from it. It is far better to cull all the animals at the time and replace them with fresh animals from a part of the rabbit unit which has not been so affected, or from another unit altogether.

It follows from this that it would be wise to consider including in the programme of the rabbit farm a practice of emptying units at particular

Disease

periods, so that the freshly-isolated parts can be easily restocked from other units where provision has been made for the new breeding stock. There are some successful farmers who consider this practice so important that they go to the point of keeping two rabbit units, alternating their occupation and then resting. This, however, is probably too extreme.

COMMON DISORDERS

Diseases are caused by *bacteria, viruses, spirochaetes, animal parasites* (including single-celled *protozoa*, various *trematodes* (flukes), *platyhelminthes* (flat worms), *nematodes* (round worms), together with mites, ticks, fleas and so on, which either produce disorders such as ear canker or spread diseases.

In addition, there is a range of general disorders arising from nutritional or physiological disturbances, nutritional deficiencies, poisons, hereditary factors and physical injuries.

There are two general points to make about diseases – firstly, disease usually arises when excessive burdens of disease organisms build up and the animal's own bodily defences cannot continue to protect it. The causes of the disease may easily be multiple and unrelated. Secondly, it is unusual that only a single condition or disease is present in an animal and quite often it is the secondary disease which causes the greatest problems.

It is uncommon for the rabbit operator to be able to diagnose the causes of problems arising in his unit. Indeed, it is true to say that quite often all the causes (or the true causes) are sometimes very difficult for even the experienced and highly-qualified person to determine. It is, therefore, essential that expert help should be sought as soon as problems are suspected. It is, however, desirable for the rabbit breeder to be aware of a few basic conditions, and their prevention and treatment.

Coccidiosis

With enteritis this, in its various forms, produces the greatest loss in terms of lowered production in the rabbit industry – not only immediately, because of the illness of the animal, but also as a result of the future lowered productivity of affected animals which never completely recover.

Coccidiosis exists in two forms – the intestinal form, produced by at least eight different species of the protozoan parasite, and the liver form,

Disease

produced by just one. The liver form is not a serious matter to the rabbit farmer, although the damaged livers may cause concern to the packer and result in a reduction in the price obtained. It is also reasonably easily prevented. The intestinal form, however, is far more important. Some of the species of intestinal coccidia are very dangerous.

Intestinal coccidiosis can only be diagnosed satisfactorily by laboratory examination of faecal pellets and by a post-mortem examination of the intestines. It is also necessary for the laboratory to be familiar with the clinical signs of the disease, which are, typically, the rapid onset of diarrhoea and loss of weight often followed by death, or recovery after a relatively short period. The problem is that these clinical signs of the disease can also be produced by other conditions (considered later).

Coccidiosis can be treated by a curative procedure with the aid of various sulpha drugs used in the drinking water. The concentrations are important and it must be done under veterinary supervision. As a general preventive measure, a coccidiostat (a chemical which prevents the development of the disease), is usually incorporated in the feeding pellets. At the present time, clopidol and robenidine are the two most commonly used. However, no coccidiostat will replace the need for the best hygiene possible in the prevention of the disease.

Diarrhoea

This is caused by a number of conditions other than coccidiosis, some of them identifiable and others not. The range of causes includes bad water, poor food (usually with a bad composition, perhaps lacking in fibre), moulds on food, some drugs, some bacteria and viruses, considerable stress to the animals in the unit and the larger intestinal parasites. Unfortunately the diarrhoea may result only some days after the cause has occurred. It is, nevertheless, important to try to establish the cause of the trouble in order to attempt to prevent it in the future.

Mention has already been made of the possibility of *intestinal parasites* causing diarrhoea. Such parasites are not common in farmed rabbits unless the animals are fed on forages and hay that have been contaminated. Occasionally, however, in such circumstances, the animals become so infected that they cause embarrassment to the processor. There are a number of proprietary anthelminthics which will eliminate intestinal parasites should the need arise.

Disease

Mucoid Enteritis

Allied to the diarrhoea, this condition was first described in the 1920s. Animals affected with mucoid enteritis adopt a very typical posture, usually suffer considerable thirst, grind their teeth in pain, lose weight rapidly and usually produce diarrhoea or almost fluid pellets which may be contaminated with mucus (a whitish thin jelly-like material). Although a great deal of research has been done on this condition, the actual causes have not been established. It is now accepted, however, that mucoid enteritis develops from an existing enteritis which has resulted from some other disorder.

Bloat

Another condition allied to diarrhoea in which the animal becomes 'blown-up' by an accumulation of gas in the abdomen. There is constipation, the same grinding of teeth as in mucoid enteritis, thirst and a lack of appetite. After a very short period the animal usually either dies or completely recovers. There is no action that a commercial farmer can take.

Respiratory Diseases

In terms of loss of profit, this is the next most damaging group of diseases in commercial rabbits, after coccidiosis and enteritis. Adults are most often affected and, although occasional mortality does occur, the diseases tend to become chronic and are passed on by the adults to the young, which are more likely to die. These diseases sometimes greatly increase in incidence and severity for no apparent reason, and in this case mortality is considerably greater.

There are several bacterial organisms involved in respiratory disorders – *pasteurella* and *bordetella* are the two most common. There is no question that the disease develops when other factors reduce the animal's susceptibility to attack, and it would be true to say that both groups of bacteria are always present in the rabbit unit to some degree.

The first signs of infection are a thin, clear discharge from the nose (which may also be produced by dust or another irritant) and this usually develops into a thick purulent discharge which may be white or yellow. The animal will wipe this sticky discharge away with its fore paws, and these will become matted as a consequence. There is constant sneezing. The condition may develop further until death occurs.

The three most important factors which aid the increase of respiratory disease, no matter what bacterial organism, or group of organisms, is involved, are nutrition, ventilation and the presence of carriers of the diseases. Therefore, the control of ventilation, the removal of dust and fumes, the prevention of rapid changes in temperature or rapid air movements, good hygiene and the elimination of any infected animals, coupled with the best possible nutrition, all contribute to a reduction in loss caused by these problems. Many attempts have been made in other countries to use vaccination against respiratory disorders, but in general, for many reasons, this has been done with very little success. Drug therapy is also of little avail.

Myxomatosis

This is the only virus disease which is of importance to the commercial rabbit farmer. It is spread most commonly in the UK by the rabbit flea, but the mosquito also acts as a carrier. The first signs are the swelling of the eyelids and other parts of the body. Death follows in most cases within about twelve days at the most. No treatment of infected animals is possible but a vaccine is available which gives excellent protection.

Prevention of myxomatosis is by way of discouraging the entrance of biting insects into the rabbit unit. In cases where the disease appears, vaccination should be immediately undertaken. Protection from the virus is acquired in a very short time (a few days only). On the Continent, where myxomatosis can be a serious problem, routine vaccinations twice a year are carried out.

Mastitis and Metritis

These two conditions of the breeding doe are a considerable problem for the rabbit farmer and occur with undue frequency. Mastitis is an inflammation of the mammary glands. It commences with hardening and swelling, the teats become painful and, after a few days, pus may be discharged. Several different bacteria are responsible, the most usual being *staphylococci*, followed by *streptococci* and *pasteurella*. The doe has little appetite but is thirsty, with a high temperature. She should be quarantined (under no circumstances should any of her young be fostered), and injections of antibiotics can be tried if the condition is found at an early stage. Treatment of advanced cases is not justified.

Metritis is an inflammation of the uterus which causes a discharge. The same bacteria that cause mastitis, together with some others (much rarer), produce the condition. Unless it is caught at an early stage, treatment tends to be unsatisfactory. The acute form of metritis can develop into a chronic form – the discharge ceases, but the animal never recovers full health and should be culled.

Disease

Spirochaetosis or 'Vent Disease'

This is a disease of rabbits which is spread by sexual contact. The sexual organs, and later the anus, become encrusted with sores which vary in size, up to half an inch (one centimetre) in diameter. The sores may run into each other to form one large mass and, if untreated, spread to lips, nostrils and eyelids. The animal remains in good condition but should never be used for mating – indeed, the desire for mating appears to be lost in such cases. The appearance of the sores occurs not less than eight weeks after infection. Prevention is by elimination of diseased animals after routine inspection on all occasions. Treatment can be carried out by veterinary injection of some antibiotics.

The Manges

These are of a serious nature and very common disorders. The one most often seen, which can cause considerable problems, is ear mange – this is produced by an invasion of the ear by one or two types of mite, *psoroptes* and *chorioptes*. The adults are about 0.02in (0.5mm) long and can be seen under a low power microscope in scrapings from the ear. The mites attack the inside of the ear and cause inflammation and intense irritation, with yellow or brown scabs being produced. The rabbit will scratch at the base of the ears and constantly shake its head. Infestation occurs by the transmission of the mites from an infected rabbit. The condition is very contagious and will spread quickly through a rabbit house. The mites can live away from the rabbit for up to at least four weeks.

The only prevention for this is regular examination of the ears of the animals and the immediate treatment of any animal that is infested. This is carried out by removing as much material as possible from the ear of the rabbit, with the aid of a small swab of cotton wool wrapped round a thin stick and soaked in hydrogen peroxide, and then applying one of proprietary ear canker insecticides, preferably an organophosphate. Initially, daily application is necessary for three or four days, and this must be followed by applications every ten days until there are no signs of the disease present. Ear canker is one of the conditions which underlines the necessity for quarantine precautions.

Body or skin mange is relatively rare. It is also caused by two species of mites (*sarcoptes* and *notoedres*) which burrow under the skin causing intense irritation, loss of fur, and scabs, with the scratching of the rabbit causing open sores. The animal does not eat and, if the condition is not treated, it will die of emaciation. It is usually best to destroy infested animals and very thoroughly disinfect the cages and equipment. If the

Disease

infection is caught in its early stages, the fur can be clipped away and a good mange preparation applied.

Sore Hocks

This is a common problem in which the pads of the hind feet (and very occasionally the forefeet) become inflamed and subsequently ulcerated. The condition appears very painful and in advanced cases the animal will lose flesh and refuse to mate or look after her young. Unless it is treated, the condition usually gets worse and a secondary infection may occur. At this stage, the whole of the pad may be affected and pus and blood discharged.

Poor cage-floor surfaces and lack of attention to hygiene are the immediate causes – rusted and damaged mesh of insufficient thickness is mostly to blame. Sore hocks occur in all varieties and weights of rabbit, but the heavier animals and those with least fur on the pad are most at risk. There appear to be some inherited factors involved in the incidence of the condition, although it may relate more to the thickness and resistance of the skin and furring on the pad.

Some relief from the condition and an aid to clearing it up is the use of a piece of clean rigid plastic sheet placed inside the cage as a temporary measure – 11in × 17in (28cm × 42cm) is a suitable size. Treatment has consisted of a cleaning of the lesions with mild disinfectants, and the use of iodine ointment or aluminium acetate solutions. Antibiotics have also been used. It has, however, recently been reported (Willis, 1986) that the use of 'Depo-Medrone' as an intra-muscular single injection at 0.1ml – 4mg/kg body weight – had excellent results in curing the condition.

Malocclusion of the Teeth

Commonly called 'buck teeth', this is a condition where the upper and lower incisors (the front teeth) do not meet. The teeth continuously grow unless worn away by working against each other. If they are not worn away, they carry on growing until they may be several inches long – this will eventually prevent the animal from eating. Although it is usually the incisors which are affected, the cheek teeth may also be misaligned and cause damage at an early stage to the tongue and cheeks. The animal suffering from the latter condition should immediately be culled, for nothing can be done. In the case of the misaligned incisors, the condition is usually caused by an inherited factor, although it may also be the result of damage to an incisor. The teeth can be clipped every three weeks or so, but it is best to eliminate the animals affected, and

Disease

to ensure that no stock is bred which may contain the factor for malocclusion in its inheritance.

Nutritional Disorders

Nutritional disorders caused by a lack of any particular constituent of food should not occur when complete pelleted diets are used. In rare cases, rickets, due to lack of calcium and phosphorous in the diet, and an absence of vitamin D (manufactured by the animal in the presence of ultraviolet light), can manifest itself. A similar condition is that of spontaneous fracture of the spine which occurs in heavily-milking does after a few litters if sufficient minerals are not included in the diet. There are other, very rare, deficiency conditions which should cause no problems, but poisoning from rat poisons, contaminated foods and so on may arise so care must be taken. Death from simple malnutrition occurs sometimes in nest young – this hapens when some condition prevents the young from suckling, or the doe from yielding milk.

7 Marketing and the Economics of Rabbit Farming

Since the start of modern intensive young rabbit meat production, the commercial rabbit industry has earned the name of 'the eighteen months industry'. There are very few activities into which so many people enter with enthusiasm and then, within two years, leave, sadly disillusioned. The reasons for this are not hard to find. Commercial rabbit farming is quite definitely not a 'get rich quick' activity. It is, in fact, probably one of the most difficult of livestock industries.

Many more people, on average, enter commercial rabbit farming without any experience or knowledge than is the case in most other professions. After they have begun there is a long learning period, for stockmanship is almost impossible to acquire solely from reading books on the subject – these can only point the way. There is also the fact that a number of people who want to work with animals do not, before they try it, really realise that animals require constant attention. No matter what happens, a routine has to be followed, and there are often unpleasant decisions to make.

One of the worst problems, however, involves the animals that are used initially to start off the enterprise. The rabbit industry has unfortunately been bedevilled by people who, realising that the price obtained for breeding stock is greater than the price for meat rabbits, and that there are always newcomers who are unable to distinguish good from bad, turn to selling breeding stock to unsuspecting and inexperienced customers.

There are first-class breeders of breeding stock, who can be trusted completely and will always give the newcomer the fairest of deals. There are others who are not like this at all. The quality of their stock (or, indeed, of the animals they have obtained for re-sale) is disastrous, although the quality, if such it can be called, of their advertising is marvellous. Many newcomers unwittingly purchase animals that could never successfully produce a profit. Nor can these animals produce future breeding stock.

The development of a profitable herd takes time. Records need to be established and future breeding stock selected from those parents which

Marketing and the Economics of Rabbit Farming

will produce a profit. It cannot be done in eighteen months. If animals are purchased at twelve weeks of age, they will not be mated for another month and it will be a further fourteen weeks before the first crop is sold. As it is not sensible, except in a particular form of farming, to purchase all the future breeding stock, a number of those first young will need to be retained. Thus, the first sales (particularly if the newcomer is so unfortunate as to buy poor quality animals), are likely to yield only a small amount of money.

The commercial rabbit industry has also been unfortunate in that some of the processors who have entered it have not been successful. This is for a number of reasons – lack of knowledge, experience and capital, for example. The rabbit farmer finds as a result that occasionally his cheque is delayed, or, in some cases, non-existent. In the worst case, the processors have not been able to accept the animals the farmer has to sell and he has then to search around for a replacement packer, often at a considerable loss.

These are the realities of the experience of many a newcomer to the rabbit industry, and some of the reasons why newly-established farmers leave the profession. Each time, another statistic is added to the population of 'the eighteen months industry'.

It is true, of course, that there are some successful rabbit farmers. Their success is due to a strict attention to detail and to the establishment of a good market.

PROCESSORS

The majority of all meat rabbits are sold live, direct to a processor. The price varies and is affected by seasonal demand – the demand is greater during the winter months when it is almost always unsatisfied. Conversely, there is often a glut during the summer months which causes problems. The larger producers can sometimes negotiate a better price for regular supplies of good and reliable quality.

Different processors, and sometimes the same processor at different times, have varying requirements in terms of the size of the young rabbits that they are looking to purchase, but in general live rabbits weighing 4.5–6lb (2–2.75kg) are wanted. Occasionally there is a demand for rather larger animals but of course the cost of production becomes higher as the size of animal increases. Culled animals are purchased by processors at prices which are usualy less than half of the price per unit of weight paid for prime young animals. At the present time prices are always quoted on a pence per pound liveweight.

Whilst many producers consider that they are not fairly treated by

Marketing and the Economics of Rabbit Farming

processors, it is also true that some producers adopt a short-sighted policy and change processors just for a small enhancement of price. This action is unwise – very often a new processor comes onto the market and, needing supplies, offers a penny or two more than the other processors, but the proportion of such people who fail for lack of experience or capital is large. If they fail, then the producer has lost his original market. It is always best to find a reliable processor and stick to him.

There is a growing tendency for some producers to market their own animals directly to the consumer. To do this some form of processing and marketing has to be set up. Farm gate sales and sales to local butchers tend to be a fairly precarious way of marketing, with demand dropping off during the summer months. The prices obtained are rather better than if the animals are sold to processors but there are expenses involved (and regulations to be obeyed) in setting up a processing unit.

The processors or packing stations obtain their supplies either with their own transport, or through collectors picking up either directly from individual farms, or at a collecting point. This is usually a farm where small producers bring their animals in order that a load of a worthwhile size may be taken away.

Only good quality meat animals of the correct weight (and that usually means at least four ounces (one hundred grams) over the minimum) should be sent for slaughter. No animal that has been on medication within a withdrawal period, no diseased or sick animals and no dirty or damaged animals should be sent.

Most processors prefer animals to have had food withheld for a period before the animal is sent, although water should never be stopped. Rabbit farmers tend to think differently as the weight is inevitably slightly reduced.

A surprising amount of bruising to the carcass can be caused if animals are roughly handled. In the same way, mixing together animals of differing ages and from different places may lead to some fighting and consequent damage, quite often with a down-grading of the meat.

Carcasses from the commercial rabbit industry are put up for sale in various forms:

1. In pieces and pre-packed with bone in.
2. As a whole carcass, usually (but not always) with head off.
3. Pre-packed, boneless fillets from the back of the animal or from the hind legs.
4. Diced boneless meat.
5. Manufactured products such as rabbit pies, etc.

Marketing and the Economics of Rabbit Farming

There is an increasing interest in the manufacture of the last three of these, as they are preferred by supermarkets – this tends to mean that the big difference between winter and summer markets is evened out.

Having considered the practices of marketing, the subject of economics must now be tackled. The profitability of rabbit farming (and by this is meant rabbit meat production without sales of anything else) depends entirely upon a satisfactory price being obtained for the young animals, and a satisfactory output of young rabbits (with low input of food), from each breeding doe cage.

The capital requirements for the setting-up of an establishment are almost invariably underestimated. They include:

1. The cost of housing, which should be based on a minimum of 15ft^2 (1.5m^2) per breeding doe cage, plus an allowance for the production of that cage.
2. The cost of cages and equipment, which, depending on the re-mate period, will include (for each breeding doe cage) cages for weaners, fatteners and a proportion for bucks.
3. Miscellaneous equipment, including a watering system, feed hoppers and lighting.
4. The initial livestock.

Some of these costs are reduced if conversion of existing buildings is undertaken, but the saving is often surprisingly small. Some idea of the capital involved is indicated by the fact that in the intensive farms in Europe, as much as twenty-five per cent of the total cost of producing meat rabbits is accounted for by servicing the capital investment, both in terms of financial charges and depreciation.

The working capital includes the total required for all expenditure (apart from capital expenditure), until returns are made by the sale of the meat animals. The major expense is food, which will probably represent seventy to seventy-five per cent of the operating costs (labour charges excluded). On this basis, and with the most favourable prices in marketing, it is necessary to have at least forty-five young per doe cage per annum actually sent to market – even then only a small profit will be made. Thereafter, every increase in the average number of young marketed per breeding doe cage means an additional profit.

Assuming the management to be satisfactory, the three key factors in the success of any rabbit farming enterprise are:

1. The actual selling price per unit liveweight.
2. The purchase price per unit weight of food.
3. The number of young sold for each breeding doe cage unit.

Marketing and the Economics of Rabbit Farming

There is a strict correlation between the cost of food per unit of weight and the selling price of meat rabbits per unit of liveweight. In the most successful enterprise (if the number of young being sold per breeding doe cage is equal), a ratio of one to eight means a satisfactory profit, if management is reasonable. A ratio of one to six (or less), unfortunately predominant in the UK even with winter prices, means that it will be extremely difficult to make a profit.

Minor reductions in food costs (providing that the food is always totally satisfactory) mean a relatively larger return; a modest increase in the average number of young sold to slaughter per breeding doe cage will mean an even greater relative profit; but the largest increase in profitability will come as the result of a relatively small increase in the sale price of the meat animals.

It is impossible to give precise figures for the various parameters, for they will all vary enormously in different circumstances and at different times. It is hoped, however, that enough has been said to enable any prospective rabbit farmer to examine the figures with which he is presented with a discerning eye, and to calculate for himself whether his circumstances are suitable to give him a chance of success.

8 Records and Recording

Records are a highly important part of the activities of a rabbit unit. Without them, the quality of management, the quality of the stock and the profit cannot satisfactorily and quickly be improved or even maintained. Records fall into three groups:

- Management records.
- Future breeding stock selection records.
- Financial records.

There are three cardinal rules in recording which must be strictly adhered to if the records are to be anything other than a waste of time.

- Records must be accurate and recorded methodically.
- Records must be as simple as possible and essential.
- To obtain their full effect records must be analysed.

The task of recording can take up a fair amount of time, therefore only those records that are absolutely necessary should be contemplated. The actual detail in the records will depend to a large extent on the activities of the rabbit unit. If, for example, the unit is concerned solely with the production of meat animals, then the future breeding stock records need not be extensive – unlike those which are required for a unit setting itself up to breed a number of lines or pure strains for sale as production stock. Some farms that profess to have such strains do not even record at all.

If the farm adopts a system (becoming more popular on the Continent) in which all breeding stock is purchased, the farm concerning itself only with the production of young animals for meat, then no records at all relating to selection need be kept.

Sometimes records can be so designed that they serve two purposes. For example, the matings register can double for the greater part of the Daily Work Sheet. There are certain purely practical points that should be considered.

The expense of drawing up a set of records, and having them well-printed on good thick sheet, is a tiny part of the total cost of recording – however, it makes the actual recording a great deal easier. An unbelievable amount of time (and, therefore, money) is wasted by

Records and Recording

having poor quality forms, by using scraps of paper which get lost and so on. The first priority is, therefore, to decide on the exact format of the records and have them drawn up and printed on good quality paper or card or, if appropriate, bound into book form.

The second point concerns the layout of the records – often insufficient space is left for the information to be written in. A line spacing of very nearly half an inch (one centimetre) is the minimum that is desirable except when it is necessary to have thirty-one lines per page (as, for example, on the monthly mortality sheet). In this case, the widest line spacing possible should be used.

Some doe (often called cage) records are printed on small cards – A5 size (148 × 210mm or 5.75 × 8.25in approx) is correct; anything less is too small. For books or full-sheet records, A4 (210 × 297mm or 8.5 × 12in approx) is the only reasonable size.

The third rule is that the facts should be accurately written down as they occur; and they should be written down in the right place the first time. The method of writing records down on bits of paper, or in a notebook, and then transferring them to the 'proper records' is not only a waste of time but leads to inaccuracies. The only partial exception to this is where the actual record sheets are used in a four ring binder (far better, incidentally, than two ring) and the sheets are taken about on a clipboard and placed in the binder afterwards. This is a very good system, but only if the sheets are numbered before use, and used in strictly consecutive order. In this way a check is kept to make sure that none are lost. The rabbit farmer must always remember that the only thing more inaccurate than the human memory when it comes to recording such information as, for example, cases of mortality in his unit, is the human mind when it makes estimates of that mortality. All records should be completed when the facts being recorded occur.

The next point concerns the protection of records (particularly the doe record card) from dirt and damp and from the ravages of the rabbits themselves. This is a matter more difficult than it may seem. A purpose-made holder, which can be fixed to the cage top, with a plastic cover is ideal but expensive. It does, however, make a good writing surface and protects the records admirably. A piece of wood at least 0.75in (2cm) larger all round than the card, with stiff transparent plastic stapled to it (under which the card can be slipped), is a good substitute.

There is then the matter of the analysis and use of the records. It may sound absurd that people will spend many hours producing the most detailed records, keeping them in excellent condition, and then will never use or analyse them, but this often happens. Unanalysed and unused records are a considerable waste of time and money.

Finally there is the question of accuracy. There is only one thing

Records and Recording

worse than no records at all and that is inaccurate records. These lead simply to the wrong results and the decisions dependent upon them are, therefore, prejudiced.

For a completely full set, those records that should be kept (apart from business or financial records) are:

1. The mating register.
2. The doe record.
3. The litter record.
4. The daily mortality record.

The analyses which should always be made are:

1. The buck performance analysis.
2. The doe comparison analysis.

The records themselves must be summarised to extract from them simple generalisations that can be used to check management.

In most of the examples of record forms given below each vertical column and each horizontal line is numbered, the first for easy reference and the second for use by the breeder.

RECORD FORMS

1. The Mating Register

This record is the main record from which much of the daily routine can originate and from which the performance of the buck is initially analysed. The page heading gives the date on which the matings are made, and then the dates on which palpation, the placing of the nest box (unless it is a permanent fixture, in which case it is the date of the nest box preparation), and the first nest box inspection are all done. This section (with the list of cage numbers in column 2) acts as a daily work register on those dates.

Columns 1, 2 and 3 are filled in with the appropriate numbers. In column 4 a tick is entered if the mating is apparently successful, or 'ref' if the doe refuses the buck. If this is the case, then the doe is listed for a future mating.

In column 5 a tick is entered, on the correct date, if the doe is found to be pregnant on palpation; 'emp' is entered if she is not. Those does found to be empty are then also listed for re-mating at a future date. A tick is placed in the next column when the nest box placement is completed.

Records and Recording

Date of mating				Date to palpate				Date to place nest box		
Date for nest box inspection										
Doe no	cage no	Buck no	Result	Palpa-tion result	Nest box place	Nest box inspection		Notes		Line
						Alive	Dead			
1	2	3	4	5	6	7	8	9	10	11
										1
										2
										3
										4
										5
										6
										7
										8
										9
										10
										11
										12
										13
										14
										15
										16
										17
										18
										19
										20
										21

Mating Register.

Records and Recording

After the first nest box inspection, columns 7 and 8 can be filled in with the appropriate numbers. The number of the doe that has kindled can now be entered elsewhere for re-mating.

Completion of the relevant section on the buck performance analysis sheet (done only *after* the final column on the mating register has been filled in) is indicated in column 10 by a tick.

2. The Doe Record

This is often called a hutch card, but this is a totally inappropriate name. It should be a full record of the doe with all the information, as each mating progresses, about her performance. In many cases, if a constant culling programme is implemented, the doe will be culled before the card is half-full, but in the case of the best does, there may be two cards. The reverse of the card is either a duplicate of the front, or is used for notes, which should be numbered, and the number entered in column 12.

Each attempted mating or presentation to the buck should be re-

Cage No:		Doe No:									
Mat-ings	Date	Buck No.	Litter			Fost-ered	Weaned				Note no
			Date	Alive	Dead	+ −	Date		Weight		
									kg	g	
1	2	3	4	5	6	7	8	9	10	11	12
1											
2											
3											
4											
5											
6											
7											
8											
Breed:			Dam no:			Sire no:					
Died/culled: date			Reason								

Doe Record.

Records and Recording

corded, no matter what the result (which should always be noted). The date of the mating, the number of the buck used and the date of subsequent kindling (if the mating is successful) are entered, as are the number of young found alive (column 5) and dead (column 6) at the first nest box inspection.

Column 7 indicates the number of young fostered to this particular doe (+) and those fostered from her (−) and columns 8–11 deal with the weaning of the young.

In column 12 the breeder can insert the number of any notes written on the reverse. This might include an analysis of the doe's mothering qualities, shown by her nest building abilities and her attitude towards her young – a tendency to neglect the litter is not a good sign.

The doe number and the cage number are given at the head of the card and the breed, dam number, sire number, the reason for removal and the date on the bottom.

3. The Litter Record

One of the most important characteristics which should be used as a basis for selection for herd improvement is the food conversion ability of the young. The only way this can be measured is by recording the food consumption and the final slaughter weight of each litter – in other words, by keeping a litter record. The doe and the litter (before and after weaning) are being fed ad lib; therefore, each amount of food placed in the hopper should be an exact number of kilos or pounds, making the additions that much easier.

On the front of the card the general details of the litter, such as parentage and weight at various times, are incorporated. On the reverse are columns for each week and each day. At the end of the week, the column is totalled and the cumulative total taken forward to the next week. Note that the litter number and cage number are given on both sides of the card, a principle which should always be followed to eliminate a possible source of error. The farmer must remember to change the cage number when the youngsters have been weaned and put into another cage.

Records and Recording

Cage Numbers

	Date	No alive	No dead		Litter no
Born					Doe no
			Total weight litter		Parity
			kg	g	
10 days					Buck no.
21 days					Breed
Weaned					
Disposal					
Notes					

Food consumption – kgs   Cage Numbers

Doe no							Litter no									
Weeks	1	2	3	4	5	6	7	8	9	10	11	12	13	14	15	16
Monday																
Tuesday																
Wednesday																
Thursday																
Friday																
Saturday																
Sunday																
Total																
B/F																
Cum total																

Litter Record Card.

Records and Recording

4. The Daily Mortality Record

Each day, any deaths discovered are totalled for each category and entered. Different vertical columns are used for the different age groups. The headings for these different columns can vary – in some cases they are more extensive. There is, however, little point in this – the categories shown are probably sufficient to give a clear idea of the actual levels of mortality and changes that occur.

Month

Date	Nest box	11-12 days	To weaning	To slaughter	Growers	Adults
1	2	3	4	5	6	7
1						
2						
3						
4						
5						
6						
7						
8						
9						
10						
11						
12						
13						
14						
15						
16						
17						
18						
19						
20						

Daily Mortality Record.

Records and Recording

The Buck Performance Analysis

The sooner a buck can be tested to determine that his performance has reached a satisfactory level, the better. Wasted cage space and food is very costly. It may be desirable to ignore the first few services of a buck when determining his performance. All the information is taken from the mating register. A single sheet is used for each buck (with continuation sheets if necessary) and each mating recorded. This can be done any time later than thirty-two or thirty-three days after the date of the mating register sheet, when the final column in the matings register will have been filled in. Details to be included in the buck performance analysis are: the date of each mating, the number of the doe served, the success or failure of the mating, and information about any young resulting (from the first nest box inspection). It is sometimes suggested that columns 7 and 8 are unnecessary, but their inclusion certainly helps to eliminate errors.

The suggestion that a small dash should be put in each square which is

Date	Doe no	Doe		Palpation		Kindled		Nest box		Line
		Mated	Refus	Doe		Yes	No	Alive	Dead	
				Preg	Empty					
1	2	3	4	5	6	7	8	9	10	11
										1
										2
										3
										4
										5
										6
										7
										8
										9
										10
										11
										12
										13
										14
										15
										16
										17

Buck Performance Analysis.

Records and Recording

not otherwise filled is a wise one. A number of people do not do this, but it probably saves time in the long run since it helps to prevent mistakes.

At any time when the breeder needs to check the performance of the buck, the ticks in each column are counted, together with the total young alive and dead. From the results a series of single figures and percentages can be obtained. For example, total satisfactory matings (or number of does pregnant), as a percentage of number mated; percentage of does refusing the buck's services; average number of young per apparently successful mating and so on.

A further extension of the buck analysis can be completed in those cases where advanced testing is undertaken, but this is outside the scope of this book. The simple ratios and percentages are used to match each buck against a standard and those bucks failing to shape up are culled.

Doe Performance Analysis

This is in two forms, depending on whether a litter record card has been completed for each litter or not. The information is taken from either the doe record card only, or that is used in conjunction with each of the litter record cards for her litters to date.

In many cases there is no need to do any more than add up the total on the doe's card. It is often immediately apparent if a doe ought to be culled. To get the full value from the records, however, a complete analysis of the cards of the best does, particularly if these are coupled with litter cards, can yield a great deal of valuable information.

Whilst all matings should be listed, it is desirable that the records over a particular period should be totalled separately and calculations made for that period. It is probably best to consider all litters from six months to fifteen months, particularly if comparisons are to be made – the time of production and the age of the rabbit at this time are important.

The information obtained from the doe record card is as follows:

1. Successful matings.
2. Refusals.
3. Litter born.
4. Number young alive at birth.
5. Number young dead at birth.
6. Number young fostered to doe (+).
7. Number young fostered from doe (−).
8. Litter weaned.
9. Number young weaned.
10. Total weaned litter weight (kg).

Records and Recording

The additional information obtained from the litter record card is:

11. Twenty-one days, number young alive.
12. Twenty-one days, total weight of litter.
13. Litter alive at disposal.
14. Disposal total number of young.
15. Disposal total weight of litter.
16. Food consumption to twenty-one days.
17. Food consumption to weaning.
18. Food consumption to disposal.

From these, a number of analyses can be completed. The most important figures that can be calculated are:

1. Conception rate.
2. Average litter size at birth.
3. Average number weaned.
4. Average litter weight at twenty-one days.
5. Average litter weight weaned.
6. Total young weaned.

Comparisons of Animals for Selection Purposes

One of the main purposes of keeping records and analysing them is so that animals can be compared and a judgement made as to which ones should be kept and which, from those retained, should be selected as the parents of the future breeding stock.

Any comparison must be made on the basis of the same characteristics at the same ages. For example, it would be unrealistic to compare the results of the first six months of a young buck with the third six months of a fully mature one. In the same way, comparing the average of the first three litters of a doe with the fourth to sixth litters would be foolish. Similarly, the food conversion rate between weaning and, say, eight weeks cannot be reasonably compared with the food conversion rate between weaning and twelve weeks – and if there has been some mortality in one group, then the total food conversion rate of that group must obviously suffer considerably in comparison with another group in which no mortality took place.

In recording food consumption and food conversion rates, the amount to one decimal place is more than sufficiently accurate – it would certainly be more accurate than the initial measurement of the food. Food consumption recording should always be started at the same

Records and Recording

period. Ideally this is when the doe is successfully mated (or re-mated, if this is the case). The overall food consumption of the whole farm is easily obtained in other, more simple ways.

There is one final point that should be made concerning weight and time scales. It is impossible to compare a weaned litter weight at, for example, twenty-eight days, with one at thirty-five days. It is, therefore, desirable, if recording of these weights is to be done, to use a standard weaning age.

Nothing has been said about financial records. Details of all expenditure and all receipts should (indeed, must, if it is a business) be kept. A simple bookkeeping primer is the best place to learn how to do this.

Some readers may be dismayed at the amount of detail given in this chapter. This is the maximum that will be needed and the information can certainly be simplified and adapted if necessary. If it is methodically done, recording does pay good dividends.

Further Reading

The following books are recommended for further reading or study. Some of these are out of print, or may become out of print – they can then often be borrowed from a library.

General

Rabbit Feeding for Meat and Fur, F.C. Aitken and W. King Wilson (Commonwealth Agricultural Bureaux, 1962)
Domestic Rabbit Biology, L.R. Arrington and K.C. Kelley (University Presses of Florida, 1976)
Rabbits' Ailments, W.P. Blount (Fur and Feather, 1957) [Out of Print]
Codes of Welfare, CRA, 1987; Ministry of Agriculture, Fisheries and Food, 1987
Rabbit Production, P.R. Cheeke, N.M. Patton, S.D. Lukefahr, J.I. McNitt (The Interstate Printers and Publishers Inc., Illinois, 6th Edition)
The Biology and Medicine of Rabbits and Rodents, J.E. Harkness and J.E. Wagner (Lea and Febiger, Philadelphia, 1983)
The Nutrition of the Commercial Rabbit, J. Lang (Commonwealth Bureau of Nutrition, Aberdeen, 1981)
The Rabbit, F. Lebas, P. Coudert, R. Rouvier, H. de Rochambeau (FAO, Animal Production and Health Series, No. 21, Rome, 1986)
The Private Life of the Rabbit, R.M. Lockley (André Deutsch, 1964)
Animal Nutrition, P. McDonald, R.A. Edwards, J.F.D. Greenhalgh (Longman, 3rd Edition)
Commercial Rabbit Production, R.J. Parkin (Ministry of Agriculture, Bulletin 50, 1985)
The Domestic Rabbit, J.C. Sandford (Collins, 1986)
A Compendium of Rabbit Production, compiled by W. Schlolaut (GTZ, Eschborn, 1985)
The Biology of the Laboratory Rabbit, edited by S.H. Weisbroth, R.E. Flatt and A.L. Kraus (New York Academic Press, 1974)

Further Reading

Genetics and Breeding

An Introduction to Animal Breeding, J.C. Bowman (Edward Arnold, 1984)

An Introduction to Practical Animal Breeding, D.C. Dalton (Granada Publishing, 1981)

Reproduction in the Rabbit, J. Hammond (Oliver and Boyd, Edinburgh, 1925)

Hammond's Farm Animals, J. Hammond Jr, J.C. Bowman, T.J. Robinson (Edward Arnold, 1983)

Animal Breeding Plans, Jay L. Lush (The Iowa State College Press, USA, 1945)

Genetic Studies of the Rabbit, R. Robinson (Bibliographia Genetica XVII, The Hague, 1957)

Colour Inheritance in Small Livestock, R. Robinson (Fur and Feather, 1978)

Appendix

USEFUL ADDRESSES

Commercial Rabbit Association
Honorary Secretary
  Mrs M Tabor
Gauntlet Chase
North Common
Sherfield English
Nr Romsey
Hampshire SO1 6JT

British Rabbit Council
The Secretary
Purefoy House
7 Kirk Gate
Newark
Nottinghamshire NG24 1AD

British Rabbit Federation
The Secretary
Hickling
Norwich
Norfolk

National Farmers' Union
Agriculture House
Knightsbridge
London SW1X 7NJ

World Rabbit Science Association
Honorary Secretary UK Branch
Dr G Partridge

Rowett Research Institute
Bucksburn
Aberdeen
Grampian
AB2 9SB

Index

Angora rabbits, 84
artificial insemination, 66
assortive breeding, 47

back-crossing, 47
bloat, 98
breeds, 39, 42
breeding
    cage, 21
    first, 66
    stock, 43
buck performance analysis, 116
building conversion, 13
bulk, 51

cages, 16
    arrangements, 20
    breeding, 21
    flat deck, 18
    installation, 23
    manufacture, 22
    requirements, 19
    sizes, 22
    space, 21
    types, 17
Californian, 39, 41
californian system, 17
capital, 106
carbohydrates, 50
cleaning, 36, 93
coccidiosis, 96
colony pens, 25
Commercial Rabbit Association, 9
conception rates, 65
conversion ratios, 53
corrective breeding, 47
culling, 74

daily mortality record, 115
development, 73

diarrhoea, 97
diseases, 89
    appearance of, 91
    build up, 95
    introduction of, 92
    spread, 92
disinfection, 93
disorders, 96
doe
    card, 109
    performance analysis, 117
    record, 112

ear tags, 77
environment control, 15
equipment, 26
    watering, 27

fats, 50
fatteners, 73
fear, 93
feed hoppers, 30
feeding
    costs, 53
    pellets, 54
    practice, 57
    quantities, 57
    systems, 55
fibre, 50
financial records, 119
first breeding, 66
flat deck cages, 23
flat deck system, 18
floors, 16
food
    additives, 54
    alternative, 55
    change, 58
    constituents, 49
    prices, 53

requirements, 52
storage, 56
values, 50
fostering, 71
fur farming, 7
fur rabbits, 85
fur removal, 37

growth, 73

handling, 59
heritability, 43
heterosis, 47
hoppers, feed, 30
housing, 10, 38
hygiene, 36, 94

identification, 76
improvement, 45
in-breeding, 47
introduction of disease, 92

killing, 80
kindling, 68
   trouble, 69

laboratory rabbits, 87
lighting, 26
line breeding, 47
litters, 70
litter record, 113
litter size, 70
live sales, 86

mains water, 28
malocclusion, 101
management, 59
   growing stock, 73
   planning, 78
   targets, 79
   time, 79
mange, 100
manufacture of cages, 22
manure disposal, 82
marketing, 9, 13
mastitis, 99
mating

assisted, 65
frequency, 65
practice, 64
register, 110
meat sales, 105
milk
   composition, 70
   production, 70
   yield, 71
minerals, 50
mucoid enteritis, 98
myxomatosis, 8, 98

nails, overgrown, 82
nest box, 32, 68
New Zealand White, 39, 40
nipple system, 27
   installation, 29
nutritional disorders, 102

out-crossing, 47

palpation, 67
pens, 16
   colony, 25
processors, 104
protein, 49
pseudo-pregnancy, 73
purchasing, 63

records, 46, 108
record forms, 110
respiratory diseases, 98
rotational cross-breeding, 47

selection comparisons, 118
selection records, 108
sexing, 63
shelters, 13
sore hocks, 75, 101
space requirements, 11
specialised pellets, 54
spirochaetosis, 100
spread of disease, 92
sterilisation, 37
stockmanship, 59
stress, 93

tattooing, 76
teeth, malocclusion of, 101
temperature, rabbit houses, 16
testing, 46
tunnel house, 12

vitamins, 50
ventilation, 15
vermin, 38

warren, 7
water
 mains, 28
 requirement, 55
watering, 27
 equipment, 27
weaning, 72
weighing, 77
wire mesh, 22

Other titles available from Crowood

A Guide to Management Series

Sheep   Edward Hart

Poultry   Carol Twinch

Ducks and Geese   Tom Bartlett

Honey Bees   Ron Brown

Practical Cheesemaking   Kathy Bliss